AF587517

# High-Current Anode Phenomena in Vacuum Arcs

**Alireza Khakpour**

born on 10.09.1981

Supervisor: Prof. Dirk Uhrlandt

Advisor: Dr. Ralf Methling

Faculty of Computer Science and Electrical Engineering
University of Rostock

This dissertation is submitted for the degree of
*Doctor Engineer (Dr. -Ing.)*

Reviewers

Dr. H. Craig Miller

Prof. Michael Kurrat

December 2018

Alireza Khakpour
High-Current Anode Phenomena in Vacuum Arcs

ISBN: 978-3-95886-254-8
1. Auflage 2018

**Bibliografische Information der Deutschen Bibliothek**
Die Deutsche Bibliothek verzeichnet diese Publikation in der Deutschen Nationalbibliografie; detaillierte bibliografische Daten sind im Internet über http://dnb.ddb.de abrufbar.



Vertrieb:

© Verlagshaus Mainz GmbH Aachen
Süsterfeldstr. 83, 52072 Aachen
Tel. 0241/87 34 34
www.Verlag-Mainz.de

Herstellung:

Druckerei Mainz GmbH Aachen
Süsterfeldstraße 83
52072 Aachen
www.DruckereiMainz.de

Satz: nach Druckvorlage des Autors
Umschlaggestaltung: Druckerei Mainz

printed in Germany

This thesis is dedicated to my lovely beautiful wife, Sona ...

# Declaration

I hereby declare that except where specific reference is made to the work of others, the contents of this dissertation are original and have not been submitted in whole or in part for consideration for any other degree or qualification in this, or any other university. This dissertation is my own work and contains nothing which is the outcome of work done in collaboration with others, except as specified in the text and Acknowledgments.

**Alireza Khakpour**
**December 2018**

# Table of contents

# Acknowledgements

That was a long uneven way to accomplish the present work which has been launched, fostered and motivated with number of people whom I am deeply grateful.

At first, I will express my deepest gratitude to Professor Dirk Uhrlandt for the encouraging, cooperative and kind way of supervising this research and giving me the great motivation to publish the results of this research in several journals. I believe that without his challenging leadership to new questions and topics the achievements of this work would not have been possible.

I am very thankful to my very kind supervisor, Dr. Ralf Methling who always supported me by constructive and fruitful discussions and unconditional effort and inspiration to explore solutions for emerging problem in the course of this work.

Many thanks go to my colleagues at Leibniz institute for plasma science and technology (INP Greifswald), Dr. Steffen Franke and Dr. Sergey Gortschakow for challenging discussions and supporting by data evaluation and preparing the optical diagnostic set-ups. I would like to thank to Gabriele Henkel and Gregor Gött for supporting during preparing the laboratory set-ups and Dr. Ruslan Kozakov for his support in data evaluation. I loved to work in this great team and it was a splendid experience to be a part of this team.

Special thanks belong to my former colleague, Mohammad Taghi Imani at Leibniz University of Hannover for interesting discussions and supports.

I very much appreciated and profited from the close cooperation and open discussion with Dr. Sergey Popov and Dr. Alexander Batrakov at Institute of High Current Electronics, Tomsk, Russia.

My special thanks appertain to Professor Steffen Großmann at Technical University of Dresden and Thomas Schönemann at ABB Switzerland who gave me the opportunity to be a part of INP Greifswald.

I am very thankful to Professor Klaus-Dieter Weltmann, the director of INP Greifswald, who gave me the opportunity to promote.

I would like to thank Dr. H. Craig Miller and Dr. Edgar Dullni at ABB Germany for very interesting discussion and providing the stimulating comments.

I would express my attitude to my family especially my parents who always have been supporting me during my study.

Finally, I express my deepest gratitude and love to my wife Sona. She always patiently relieved my daily strain and helped me to be still motivated. This work could not be appeared without her support.

This work is supported by Leibniz Institute for Plasma Science and Technology (INP Greifswald).

# List of own publications

Several journals and conference contributions results as an outcome of this thesis which are integrated in the text and indicated with a star (*).

## Journals

1. **A. Khakpour**, St. Franke, D. Uhrlandt, S. Gorchakov, and R. Methling. Electrical arc model based on physical parameters and power calculation. *IEEE Trans. Plasma Sci.*, 43(8):2271-2729, 2015.*

2. **A. Khakpour**, St. Franke, S. Gortschakow, D. Uhrlandt, R. Methling, and K. D. Weltmann. Electrical arc model based on the arc diameter. *IEEE Trans. Power Del.*, 31(3):1335-1341, 2016.*

3. **A. Khakpour**, D. Uhrlandt, R. Methling, S. Gortschakow, St. Franke, M. T. Imani, and K. D. Weltmann. Impact of temperature changing on voltage and power of an electric arc. *Electr. Power Sys. Res.*, 143, 73-83, 2017.*

4. **A. Khakpour**, D. Uhrlandt, R. Methling, St. Franke, S. Gortschakow, S. A. Popov, A. V. Batrakov, and K. D. Weltmannn. Impact of different vacuum interrupter properties on high-current anode phenomena. *IEEE Trans. Plasma Sci.*, 44(12):3337-3345, 2016.*

5. **A. Khakpour**, S. Gortschakow, D. Uhrlandt, R. Methling, St. Franke, S. A. Popov, A. V. Batrakov, and K. D. Weltmann. Video spectroscopy of vacuum arcs during transition between different high-current anode modes. *IEEE Trans. Plasma Sci.*, 44(10):2462–2469, 2016.*

6. **A. Khakpour**, St. Franke, R. Methling, D. Uhrlandt, S. Gortschakow, S. A. Popov, A. V. Batrakov, and K. D. Weltmann. Optical and electrical investigation of transition from anode spot type 1 to anode spot type 2. *IEEE Trans. Plasma Sci.*, 45(8):2126-2134, 2017.*

7. **A. Khakpour**, R. Methling, D. Uhrlandt, St. Franke, S. Gortschakow, S. A. Popov, A. V. Batrakov, and K. D. Weltmann. Time and space resolved spectroscopic investigation during anode plume formation in high-current vacuum arc. *J. Phys. D: Appl. Phys*, 50(18):185203, 2017.*

8. **A. Khakpour**, S. Popov, St. Franke, R. Kozakov, R. Methling, D. Uhrlandt, and S. Gortschakow. Determination of Cr density after current zero in a high-current vacuum arc considering anode plume. *IEEE Trans. Plasma Sci.*, 45(8):2108-2114, 2017.*

9. **A. Khakpour**, M. T. Imani, R. Methling, St. Franke, S. Gortschakow, and D. Uhrlandt. An improved electric arc model for vacuum arc regarding anode spot modes. *IEEE Trans. Dielectr. Electr. Insul.*, accepted.

10. **A. Khakpour**, R. Methling, St. Franke, S. Gortschakow, and D. Uhrlandt. Vapor density and electron density determination during high-current anode phenomena in vacuum arcs. *J. Appl. Phys.*, submitted.

11. **A. Khakpour**, R. Methling, St. Franke, S. Gortschakow, and D. Uhrlandt. Emission spectroscopy during high-current anode modes in vacuum Arc. *J. Plasma Phys. and Tech.*, 4(3), 2017.*

12. M. V. Lisnyak, A. V. Pipa, S. Gorchakov, S. Iseni, St. Franke, **A. Khakpour**, R. Methling, and K. D. Weltmann. Overview spectra and axial distribution of spectral line intensities in a high-current vacuum arc with Cu-Cr electrodes. *J. Appl. Phys.*, 118(12):123304, 2015.*

13. R. Methling, S. Gorchakov, M. V.Lisnyak, St. Franke, **A. Khakpour**, S. A. Popov, A. V. Batrakov, D. Uhrlandt, and K. D. Weltmann. Spectroscopic investigation of a Cu-Cr vacuum arc. *IEEE Trans. Plasma Sci.*, 43(8):2303-2309, 2015.*

14. S. Gortschakow, **A. Khakpour**, S. Popov, St. Franke, R. Methling, and D. Uhrlandt. Determination of Cr Density in a High-Current Vacuum Arc Considering Anode Activity. *J. Plasma Phys. Tech.*, 4(2):190-193, 2017.*

15. R. Kozakov, **A. Khakpour**, S. Gorchakov, D. Uhrlandt, D. Ivanov, I. Murashov, G. Podporkin, and V. Frolov. Investigation of a multi-chamber system for lightning protection at overhead power lines. *J. Plasma Phys. Tech.*, 2(2):150-154, 2015.

16. D. Uhrlandt, R. Methling, St. Franke, S. Gorchakov, M. Baeva, **A. Khakpour**, V. Brüser. Extended methods of emission spectroscopy for the analysis of arc dynamics and arc interaction with walls. *J. Plasma Phys. Tech.*, 2(3):280-289, 2015.

17. R. Methling, **A. Khakpour**, S. Wetzeler, D. Uhrlandt. Investigation of an ablation-dominated arc in a model chamber by optical emission spectroscopy. *J. Plasma Phys. Tech.*, 4(2):153-156, 2017.

18. Y. Firouz, S. Farhadkhani, J. Lobry, F. Vallée, **A. Khakpour**, O. Durieux. Numerical comparison of the effects of different types of distributed generation units on overcurrent protection systems in MV distribution grids. *J. Renewable Energy*, 69, 271-283, 2014.

19. M. Amiri, **A. Khakpour**, A. Golshani. Economic evaluation of CHP units installation in residential buildings of Iran in case of energy subsides removal. *J. Life Sci.*, 10(8s):376-380, 2013.

## Conference proceedings

1. **A. Khakpour**, A. Golshani, Th. Schoenemann, S. Gorchakov. Transient analysis of Tehran power grid considering planned expansion model. In *IEEE Energy Conf. (ENERGYCON)*, pages 291-298, Cavtat, Croatia, 2014.

2. A. Golshani, **A. Khakpour**, and L. Hofmann. FACTs based fault Current limiter utilization in HV substation. In *25th IEEE Canadian Conf. Elec. and Comp. Eng. (CCECE)*, Montreal, Canada, 2012.

3. **A. Khakpour**, A. Golshani, M. Farahani. Determining the Possibility of Paralleling Generators with Different Grounding. In *Proc. 46th Int. Uni. Power Eng. Conf. (UPEC)*, Soest, Germany, 2011.

4. **A. Khakpour**, S. Gortschakow, S. A. Popov, R. Methling, St. Franke, D. Uhrlandt, A. V. Batrakov, and K. D. Weltmann. Time and space resolved video spectroscopy of the vacuum arc during the formation of high-current anode mode. In *Proc. 27th Int. Symp. Discharges Elect. Insul. Vac.*, pages 287-290, Suzhou, China, 2016.*

5. **A. Khakpour**. Impact of AC and pulsed DC interrupting currents on the formation of high-current anode modes in vacuum. In *Proc. 27th Int. Symp. Discharges Elect. Insul. Vac.*, pages 129-132, Suzhou, China, 2016.*

6. S. A. Popov, A. V. Schneider, V. Lavrinovich, A. V. Batrakov, S. Gortschakow, and **A. Khakpour**. Fast video registration of transition processes from diffuse mode to anode spot mode in high-current arc with copper-chromium electrodes. In *Proc. 27th Int. Symp. Discharges Elect. Insul. Vac.*, pages 375-378, Suzhou, China, 2016.

7. R. Methling, S. Gorchakov, M. V. Lisnyak, St. Franke, **A. Khakpour**, S. A. Popov, A. V. Batrakov, D. Uhrlandt, and K. D. Weltmann. Spectroscopic investigation of high-current vacuum arcs. In *Proc. 26th Int. Symp. Discharges Elect. Insul. Vac.*, pages 221-224, Mumbai, India, 2014.*

8. St. Franke, R. Kozakov, S. Gortschakow, **A. Khakpour**, R. Methling, and D. Uhrlandt. Broadband absorption technique applied to a free-burning arc in ambient air. In *21th Int. Conf. on Gas Discharges and their Application*, pages 5-8, Nagoya, Japan, 2016.

9. J. Pettersson, M. Becerra, St. Franke, S. Gortschakow, **A. Khakpour**, and R. Bianchetti. Space-resolved spectroscopic and photographic studies of the vapor layer produced by arc-induced ablation of polymers. In *21th Int. Conf. on Gas Discharges and their Application*, pages 149-152, Nagoya, Japan, 2016.

10. St. Franke, R. Methling, **A. Khakpour**, S. Gorchakov, V. Brüser, D. Uhrlandt. Study of high-current arcs and its interaction with side walls and layers. In *24nd IEEE Int. Conf. Plasma Sci.*, Antalya, Turkey, 2015.

11. St. Franke, M. Lisnayk, S. Gorchakov, **A. Khakpour**, R. Methling, A. Pipa, D. Uhrlandt, and K. D. Weltmann. Spatial distribution of charged particle emission in

a copper-chromium high-current vacuum arc. In *24nd IEEE Int. Conf. Plasma Sci.*, Antalya, Turkey, 2015.

12. D. Uhrlandt, S. Gorchakov, V. Brueser, St. Franke, **A. Khakpour**, M. Lisnyak, R. Methling, Th. Schoenemann. Interaction of a free burning arc with regenerative protective layers. In *J. Phys.: Conf. Ser.*, 550(1):012010, 2014.

# Abstract

The emerging interest on the application of vacuum interrupter (VI) in medium and even high-voltage application has led to increase of comprehensive and fundamental research on VIs. Application of these interrupters is however hindered by limitation of interrupting capability, when high-current anode phenomena occur.

The goal of the work is to investigate the high-current anode phenomena using spectroscopic methods se well as electrical measurements. Different current waveforms, i.e. AC and DC pulses are applied. Moreover, the impact of vacuum interrupter properties, e.g. electrode diameter and geometries, contact materials, opening time and opening speed on the formation of high-current anode phenomena are investigated.

The results reveal that the threshold current of the high-current anode formation is dominated by the current in case of high-frequency pulsed, whereas in case of pulsed DC, it is controlled by transferred charge.

Along with typical and already known high-current modes, i.e. footpoint, anode spot, and intense modes, a new type of high current anode mode, anode spot type 2, is introduced. Anode spot type 2 is similar to the intense mode but it appears at larger gap length. In contrast to anode spot (anode spot type 1), both anode and cathode are active in case of anode spot type 2. The line emission intensities of all species increase near the anode and the cathode in case of anode spot type 2. The anode spot type 2 is examined by determining radiating density, ground state density and electron density which is about two times higher compared to anode spot type 1.

Optical emission spectroscopy combined with high-speed camera imaging (video spectroscopy) is used to investigate the temporal and spatial distribution of Cu I, Cu II, and Cu III during discharge modes. The results show different patterns during discharge modes.

The formation of anode plume, is investigated using video spectroscopy. Existence diagrams confirm that the anode plume appears always after the extinction of anode spot type 2 in case of CuCr electrodes before current zero. The results unfold that the emission from the inner part of anode plume is covered by atomic lines, whereas the outer part is dominated by ionic line radiation. Anode plume formation which is a result of shock effect in front of the anode shows that anode plume contracts first and then it expands. Radiating densities of atomic and ionic Cu lines, and temporal evaluation of temperatures inside the anode plume are determined during the formation of anode plume.

The ground state density of Cr I before and after current zero is calculated for two cases; with and without anode plume using broadband absorption spectroscopy. The results point out that the ground state density after current zero is higher in case of anode plume. Moreover, during active phase the ground state density is higher in case of anode spot type 2 compared to anode spot type 1. Furthermore, the temporal evolution of ground state density of Cr I is calculated during high-current anode phenomena.

A new electric arc model based on the existence diagram is proposed in this work. In contrast to conventional electric arc model, e.g. Mayr, Cassie, Habedank, Schavemaker, etc., this model can track or detect the arc voltage of vacuum arc even during anode spot type 2 and abrupt change in the arc voltage. Moreover, this model based on the existence diagram has a predictability characteristic. By providing the existence diagram to the model, the arc voltage can be traced in quite different interrupting current ranges.

# Zusammenfassung

Das große Interesse an der kommerziellen Anwendung von Vakuumschaltern im Mittelspannungs- und in zunehmendem Maße auch im Hochspannungsbereich führte zu umfassenden und grundlegenden Forschungen an diesem Betriebsmittel. Eine Erweiterung des Anwendungsbereichs der Vakuumschalter ist abhängig von bestimmten Anodenphänomenen, welche bei hohen Strömen auftreten und die Fähigkeit zur Stromunterbrechung limitieren.

Ziel der Arbeit ist die Untersuchung dieser Hochstrom-Anodenmodi mit optischen, insbesondere spektroskopischen Methoden sowie mit elektrischen Messungen. Um die zeitliche und räumliche Verteilung der Linienstrahlung von Kuperatomen (Cu I) sowie von einfach (Cu II) und doppelt geladenen Ionen (Cu III) zu untersuchen, wird optische Emissionsspektroskopie in Kombination mit einer Hochgeschwindigkeitsvideokamera verwendet (Videospektroskopie). Es werden verschiedene Stromformen angewandt, insbesondere Wechselstrom- und Gleichstromimpulse unterschiedlicher Dauer. Darüber hinaus wird der Einfluss diverser Spezifikationen von Vakuumschaltern, wie z.B. Elektrodengeometrien, Kontaktmaterialien, Lichtbogendauer und Öffnungsgeschwindigkeit, auf die Entstehung der Hochstrom-Anodenmodi erforscht.

Anhand der Untersuchungen wird nachgewiesen, dass die Schwelle für die Ausbildung der Anodenphänomene bei Wechselspannungsimpulsen durch den Strom dominiert wird, während sie bei Gleichspannungsimpulsen durch die übertragene Ladung bestimmt wird.

Neben bereits bekannten Hochstrommoden, wie dem Fußpunkt, dem Anodenspot und dem intensiven Mode, wird ein neuer Typ erstmals beobachtet und beschrieben, der als Anodenspot Typ 2 bezeichnet wird. Er ähnelt dem intensiven Mode, erscheint jedoch bei viel größeren Elektrodenabständen. Im Gegensatz zum bereits bekannten Anodenspot (nunmehr als Typ 1 bezeichnet) sind beim

Anodenspot Typ 2 sowohl die Anode als auch die Kathode aktiv. Die Emission von Spektrallinien aller Spezies erhöht sich im Falle des Anodenspots Typ 2 in der Nähe der Anode und der Kathode. Der Anodenspot Typ 2 wurde durch Ermittlung der Strahlungsdichte, der Grundzustandsdichte und der Elektronendichte, die im Vergleich zum Anodenspot Typ 1 ca. zweimal größer ist, untersucht.

Die Entstehung der erst kürzlich entdeckten Anodenplume [1] kurz vor dem Stromnulldurchgang wurde für CuCr-Elektroden mit Hilfe der Videospektroskopie herausgearbeitet. Existenzdiagramme deuten darauf hin, dass die Anodenplume immer nach dem Auslöschen des Anodenspot Typ 2 erscheint. Anhand von Versuchsreihen hat sich herausgestellt, dass die Emission aus dem inneren Teil der Anodenplume weitgehend durch Atomlinien generiert wird, während im Randbereich die Strahlung einer höheren Anteil von Ionenlinien aufweist. Die Anodenplume zieht sich mit sinkendem Strom zunächst zusammen und dehnt sich kurz vor Stromnull wieder aus. Die Strahlerdichten von Atom- und Ionenlinien von Cu und der zeitliche Temperaturverlauf innerhalb der Anodenplume wurden ebenfalls bestimmt.

Die Grundzustandsdichte von Cr I wurde mittels optischer Absorptionsspektroskopie gemessen. Sie ist sowohl vor als auch nach dem Stromnulldurchgang im Fall der Anodenplume höher. Darüber hinaus wurde die zeitliche Entwicklung der Grundzustandsdichte von Cr I während der Hochstrom-Anodenmodi berechnet.

Zur Simulation von Vakuumbogentladungen mit Anodenphänomenen wird ein neues Modell vorgestellt, welches auf dem Existenzdiagramm basiert. Im Gegensatz zu herkömmlichen Lichtbogenmodellen, z.B. nach Mayr, Cassie, Habedank oder Schavemaker, kann dieses Modell den realen Verlauf der Lichtbogenspannung deutlich besser wiedergeben. Insbesondere die sprunghafte Änderung während des Übergangs zum Anodenspots Typ 2 und der Anodenplume kann qualitativ und quantitativ nachvollzogen werden. Unter der Voraussetzung eines zuvor bestimmten Existenzdiagramms besitzt das Modell eine Vorhersagbarkeitseigenschaft, so dass die Lichtbogenspannung für verschiedene Unterbrechungsströme berechnet werden kann.

[1] plume = Abluftfahne, Rauchschwaden

# Chapter 1

# Background and motivation

Nowadays vacuum interrupters (VIs) are widely employed in medium voltage level due to their privileged electric and dielectric properties together with their economic and environmental benefits [1–3]. However, application of vacuum interrupters in high-voltage (>100 kV) is limited due to some technical and economic issues such as electron field emission, thermal and dielectric recovery, costs, etc [4–6].

A vacuum interrupter reaches its interruption limit once high-current anode phenomena occur due to significant melting and evaporation of anode surface [4, 5]. This can lead to pressure establishment of neutral metal particles. A part of the metal particles can be ionized. Nevertheless, a metal vapor which exists even after current zero can reach the critical values and transient recovery voltage (TRV) leads to a breakdown and failure of vacuum interrupter [4–6]. Different strategies are followed to prevent vacuum interrupters from early failure, e.g. applying axial magnetic field (AMF) or transverse magnetic field (TMF) [2]. In the case of TMF contact systems, the electrode geometry introduces a tangential magnetic field into the vacuum arc urging a usually constricted arc to move azimuthally on the electrode surface. In the matter of the AMF electrode design, an axial magnetic field is generated, which allows vacuum interrupter to operate in diffuse mode [3, 6]. In diffuse mode, the cathode plays an active role in the discharge process, while the anode is passive: collecting particles emitted by the cathode.

Different high-current anode modes have been classified regarding voltage characteristics, illumination area near the electrodes and electrode surface temperature, i.e. footpoint, anode spot, and intense mode, where the anode plays an active role [7–12].

Regarding the footpoint mode, one or more small bright spots appear on the anode. Meanwhile the gap between electrodes is mostly filled with a fairly bright diffuse glow similar to diffuse arc mode [8]. In this case, the arc voltage is relatively high and noisy. The anode spot mode is a high-current mode with considerable anode activity in which one large or several smaller and very bright spots are present on the anode. Transition to the anode spot mode is accompanied typically by an abrupt change in the arc voltage, which is relatively high and normally without noise in this mode [8]. In the intense mode, large spots occur at both electrodes, which are brighter than those in the anode spot mode. The intense arc mode normally occurs at lower electrode distances compared with the anode spot [9]. The arc voltage remains low and without noise but is still higher than that in the diffuse mode. However, the anodic arc may also exist at relatively low arc currents and lower anode surface temperatures, where the evaporation may occur over the whole anode surface. This mode of the anodic arc is known as the hot-anode vacuum arc or HAVA [8].

The transition from diffuse mode to high-current anode mode has been investigated widely by means of evaluation of images captured by high speed camera beside electrical characteristics [13, 14].

Formation of aforementioned modes is controlled by electrode distance, instantaneous current, transferred charge, electrode material, electrode diameter, magnetic forces, and current waveform as well. Existence diagrams of the formation of different discharge modes, i.e. gap length vs. arc current has been derived to investigate the influence of different parameters which are valid only for a set of certain system parameters [8–10].

For these modes, the surface temperature can extremely exceed the melting point of the electrode material and hence it gives rise to excessive electrode evaporation. The high-current anode modes are expected to strongly affect the dielectric and thermal recovery [15, 16]. Consequently, the circuit breaker lifetime is affected resulting in electrode erosion.

The high-current modes have been the subject of numerous studies by means of optical emission spectroscopy during the active phase. Nevertheless, the transition between different discharge modes with respect to spectroscopic information about atomic and ionic lines is concerned in few [17–19]. The intensity of different Cu lines was investigated during formation of the HAVA mode in a DC arc [17]. A spectroscopic analysis is conducted at three positions, close to the anode and the cathode and in the middle of the gap between Cu-Al electrodes. However, the higher ionization states of copper which may be important for the arc dynamics could not be readily observed. Some spectroscopy analyses are also conducted to extract information regarding the axial distribution of atomic and ionic copper lines but in diffuse mode [20, 21].

Dielectric recovery of vacuum interrupters after high-current faults is scrutinized in few works. In this regard, the Cu density is determined during and after interruption of high-current vacuum arc using laser absorption techniques [22, 23]. In fact, presence of metal vapor or plasma after current zero can reduce the dielectric strength leading to reignition or even breakdown.

Notably, there is still a lack of research regarding spectroscopic analysis of high-current anode phenomena. By applying video spectroscopy with high temporal resolution more information concerning atomic and ionic particle density distribution, plasma temperature, and mass distribution could be extracted.

Prior, during, and after current zero, in which light emitted from the plasma is very week, absorption spectroscopy can be applied to get more information with respect to the evaporation of electrode material resulting in recovery failure.

There still remains a number of open questions related to the mechanisms leading to the formation of high-current anode modes. It is unsatisfactory that up to now it is just possible to describe anode modes and their occurrence but not being able to physically motivate their formation.

In addition, electric arc models of the high-current vacuum arc might be useful. such a model should include also the change of arc voltage during high-current anode modes which can be probably deduced from the existence areas.

In this work, after background and motivation section, a brief introduction regarding vacuum interrupters and its application is presented in chapter 2. The experimental setup together with the applied methods specifically optical and electrical diagnostics, is presented in chapter 3. In chapter 4, the impact of different technical parameters of vacuum interrupters on the formation of high-current modes are explained: waveforms, opening time, contact material, geometry, and contact opening speed. In this chapter, a new type of high-current anode modes, anode spot type 2, is introduced. Chapter 5 comprises investigation on the transition between the different high-current modes by the video spectroscopy. In this chapter, some physical information, e.g. vapor density, radiating density, and electron density is also provided. Chapter 6 introduces another new type of high-current mode (anode plume) which appears just before current zero crossing and is characterized by optical emission spectroscopy. In this study, particle density during different high-current anode modes and after current zero is determined by broadband absorption spectroscopy technique, see chapter 7. In chapter 8, an electrical arc model based on the existence diagram is proposed. The obtained results and findings are discussed in chapter 9 followed by the summary and outlook in chapter 10.

# Chapter 2

# Introduction to vacuum interrupters

## 2.1 Application of vacuum interrupter

High-current vacuum interrupters have found wide applicability in medium voltage due to the excellent electric and dielectric properties beside consequent economic and environmental advantages. There is also great interest to develop circuit breakers based on vacuum technology for high voltages. Nonetheless, application of vacuum interrupters in high voltage field is impeded due to the technical difficulties, e.g. thermal and dielectric recovery as well as economic issues i.e. high manufacturing costs. This underlines the importance of further research in this field to address those aforementioned issues [1, 3].

In fact, appearance of high-current anode modes goes along with temperature increase of anode surface. This leads to vaporization of metal particles from the contact surface which occupy the inter-electrode area [24–26]. Thereby, the dielectric withstand strength of the vacuum gap is reduced and the dielectric recovery strength after current zero-crossing is deteriorated. In this regard, the interruption limit of a vacuum switch is often reached by appearance of the high-current anode phenomena [4, 5, 27].

The main specifications of vacuum interrupters to be preferred in medium voltage along with promising service experiences are as follows:

- Excellent dielectric withstand strength.

- Swift interruption performance and consequently appropriate for fault clearance.
- Robust against frequent operation.
- Explosion-free and without fire hazard.
- Quiet operation and almost maintenance free.
- Long life span and lightweight.

The type of the load determines the switching performance of the vacuum interrupters: high power factor loads, inductive loads, and capacitive loads.

Interruption of the high power factor loads (quasi-ohmic) is carried out when the current and the voltage are almost in the same phase. Hence, the network voltage will be approximately zero when the load current approaches natural zero. Thereby, transient recovery voltage (TRV) will be as small as that disconnection occurs without any critical stress on the vacuum interrupter.

In comparison, maximum phase shift between current and voltage happens in the case of inductive loads. Therefore, when current approaches zero the voltage is close to the maximum value, which may lead to the re-ignition. Voltage fluctuations across the electrode contact appears as a result of the interaction of grading capacitance and the system inductance [28]. The rate of rise of transient recovery voltage (RRRV) by interruption of inductive loads is higher compared to the interruption of resistive or capacitive loads. To ensure a reliable switching, dielectric recovery of the contact gap after current interruption must be faster than rate of rise of TRV imposed by the connected system; if not, reignition is unavoidable. The usual high current inductive loads, namely, unload transformers, shunt reactors, and arc furnaces, huge motors and short circuits being supplied with current of a few amperes, hundreds of amperes, and kilos of amperes, respectively, impose stresses, which could be endured by VCBs because of their rapid dielectric recovery properties [1, 28].

Capacitive voltage elements as high voltage cables or transmission lines exhibit a low capacitance compared to the capacitor banks. Therefore, energization of capacitor banks is accompanied by much higher inrush current than cables or lines due to the low surge impedance of capacitor banks resulting in thermal energy input for pre-strike. Consequently, energization of high capacitive loads arises more challenges for vacuum switches [1, 28].

The inrush current imposed by pre-strike brings about local over-heating of the electrode surfaces and consequently welding of the contacts. This initiates metal vaporization of the contact surface and micro protrusions which can in turn deteriorate the dielectric withstand strength of the contact systems. This condition will be worse when the closing operation is followed by an opening with very small current [1, 28].

## 2.2 Breakdown in vacuum

Vacuum switches should necessarily have the property of fast dielectric recovery before current zero. After current switching and arc extinction, hot post arc plasma could be still observed in the gap. This post-arc plasma has to be disappeared by the mechanism of rapid diffusion. Commonly, the interruption process is carried out in two critical steps: thermal and dielectric recovery [3, 6].

**Thermal recovery**

Once the current approaches zero, the residual discharge should be disappeared before TRV increases to the value high enough for thermal reignition. By occurrence of fast rising of TRV or high RRRV, the thermal recovery property plays a vital role in flawless interruption process. These stresses could be seen when faults on overhead lines near the circuit breaker happen or when the circuit breaker is integrated close to the CLR (core less reactor) [28].

Before current zero, there is no interaction between the network and the vacuum switch. Due to the very low and constant arc, the arc current reaches zero virtually uninfluenced and with a maximum di/dt.

Vacuum interrupter can switch those very steep currents up to many hundreds of A/$\mu$s. The very efficient thermal recovery property allows the circuit breakers to withstand very high RRRV [28].

**Dielectric recovery**

About hundreds of microseconds after current zero-crossing, the major part of the discharge is quenched. The generated TRV could impede the extinction of the arc by imposing an additional stress across the gap in which the vaporized metal particles are presented because of arcing time. In the case of a short circuit, the minimum arcing time is usually short. Hence, the vacuum switches should be designed to withstand those stresses. Vacuum interrupters are equipped with just a single contact pair, which is not the case for SF6 switches. This contact pair must withstand the power frequency recovery voltage and the TRV overvoltages as well. In fact, surface roughness or in other words the surface microstructure has a major influence on breakdown voltage in vacuum switches [6].

It should be noted that the arcs can play a positive role where they can burn away the metal whiskers and protrusions being present over the contact surface, which initiate and develop the breakdown process. In contrast, considering low current switching by capacitive or inductive loads, the arc energy is not as high as to melt down and eradicate those sharp tips on the contact surface. Hereupon, late breakdown is detected quite often after TRV climax in low current vacuum interrupters [28].

## 2.3 Transient recovery voltage

Generally, recovery voltage is addressed to those voltages which appear across the terminal of the switch, once the current is interrupted. Meanwhile, conductivity of the discharge drops. After current zero, the post arc current flows with a value ranging from10 to 20 A for a few microseconds due to residual charge, accordingly, TRV generated.

TRV is generated as switching response of the network connected to the switch by interruption of the current. TRV could also be defined as voltage difference between the supply side and the load side of the interrupter. In other words, TRV is influenced by the type of the network element connected to the interrupter; capacitive, resistive, inductive, or a combination of them [29].

The interruption process can be successfully carried out if the interrupter can withstand the common emerging stresses as TRV, RRRV, and the power frequency recovery voltage. TRV can exhibit either oscillatory, triangular or exponential behavior. The worst case scenario occurs across the first pole to open of a circuit breaker by clearing a three-phase symmetrical fault at its terminal when the system voltage is maximum. The oscillatory TRV is generated when the current of the fault flows passes through an inductive element, e.g. a transformer and reactor, and transmission line or cable as damping component is not presented (Fig. 2.1) [29].

An exponential TRV as depicted in Fig. 2.2 appears after the clearance of a three-phase fault at the breaker terminals, if at least one transformer and one line are present on the fault free side of the switch [29].

Fig. 2.3 represents the topology of the system and TRV behavior in the case of triangular recovery voltages. The steepness of the TRV is attributed to the fault location; it increases more as the fault occurs closer to the interrupter [29].

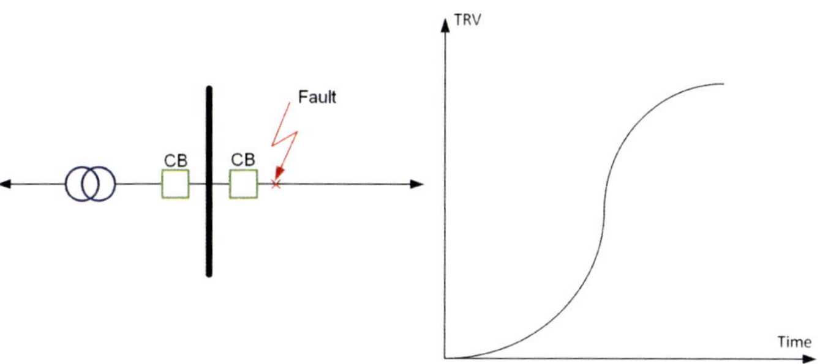


**Fig. 2.1** Oscillatory TRV characteristic.

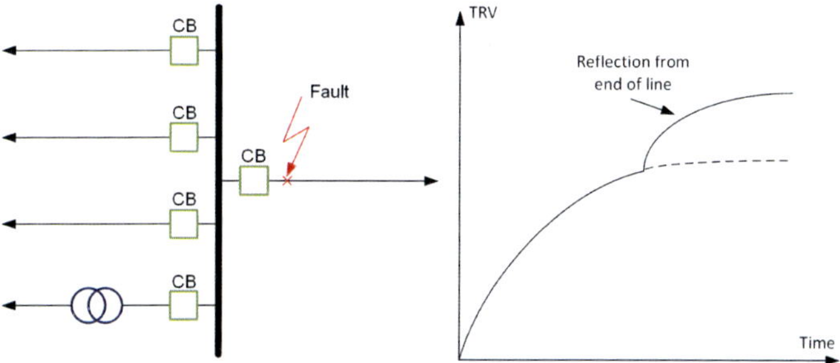


**Fig. 2.2** Exponential TRV characteristic.

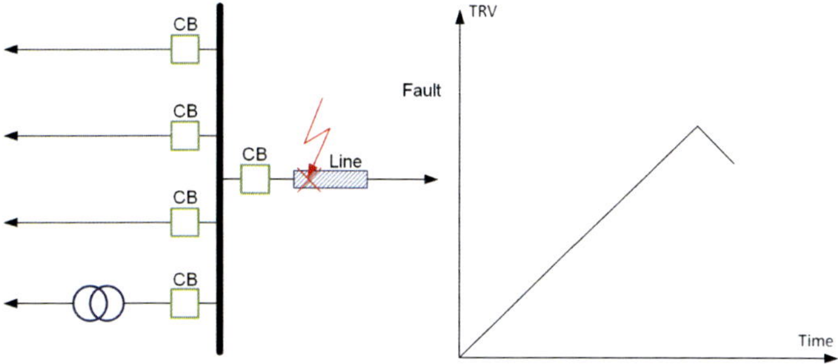


**Fig. 2.3** Short-line fault TRV characteristic.

The behavior of the TRV is dependent on the line surge impedance $Z$. The rate of increase of the saw-tooth shaped TRV is usually higher than that observed by exponential or oscillatory TRVs. It should be noted that vacuum interrupters could regain the dielectric recovery property within microseconds after the fault current. Accordingly, the vacuum based breakers are preferred for clearance of the short line faults (SLF) due to the ideal switching performance [29].

## 2.4 Methodology of vacuum arc control

Construction of a vacuum interrupter is schematically illustrated in Fig. 2.4. Current interruption is realized by separation of the moving contact from the fixed one. Meanwhile, plasma is formed in the inter-electrode area. This is accompanied usually by formation of arc as a diffuse mode in the case of low current. The cathode spots can usually be observed resulting in cathode erosion but without overheating of the cathode surface. Erosion of the anode is not detected.

1 Fixed contact stem
2 Fixed terminal pad
3 Ceramic insulator
4 Fixed contact
5 Metal shield
6 Moving contact
7 Insulator
8 Bellows
9 Bearing
10 Moving contact stem
11 Mechanical coupling for operating mechanism

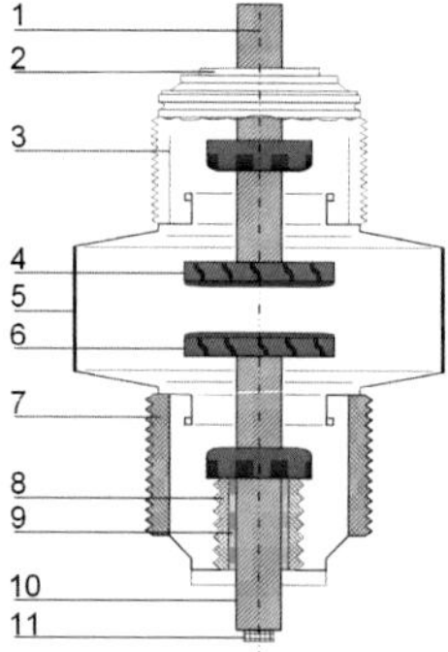


**Fig. 2.4** The schematic representation of vacuum interrupter.

Surface of the anode is overheated locally because of high current that results in contact erosion. It is of critical importance that the discharge is homogeneously distributed over the contact surface in order to relieve the resulting erosion. In this regard, two mechanisms are introduced and commonly applied to control the discharge in vacuum interrupter, namely the application of either a radial or transverse magnetic field (RMF or TMF) or of an axial magnetic field (AMF) [1, 3, 6, 30]. Using TMF design, the arc is confined because of pinch effect. Thereby, the discharge is pushed over the contact surface as a result of Lorenz force. In vacuum interrupters equipped with AMF contact, however, the position of the discharge remains unchanged and due to the magneto-dynamic effect the arc is spreads over the contact surface. AMF and TMF drive the discharge to transform to diffuse mode or multiply arcs at the end of a current half wave, respectively. TMF contacts have rather simple design. Accordingly they are cost effective in design and thus are preferred for medium voltage application. AMF technology is commonly used for higher voltage and for switching of capacitor banks.

**TMF contacts**

The typical TMF contact configuration is illustrated in Fig. 2.5. The current flows through contacts resulting in a magnetic field component perpendicular to the arc. This is also termed RMF since the direction of magnetic field is as the same with radial direction. The strength of TMF is determined by the arc current and in the order of 10 mT/kA [6]. In TMF contact, arc remains pillar-shaped at high currents and moves over the electrode surfaces. As shown in Fig. 2.5, simultaneous shear and Lorentz forces result in inward rotation with opposite direction. The Lorentz-force resulting from discharge current and TMF effect propel the arc in direction perpendicular to current and TMF [1, 3, 6].

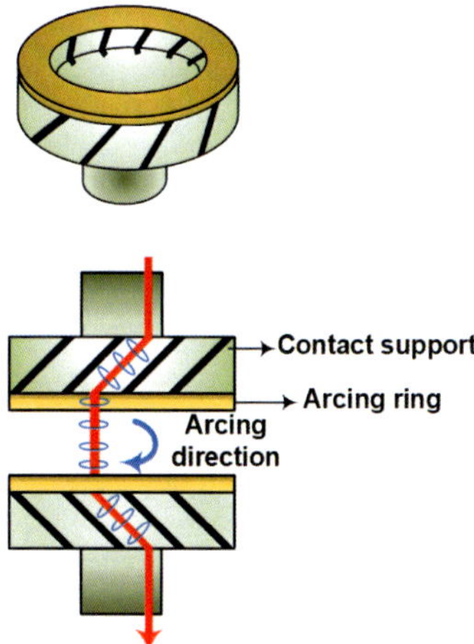


**Fig. 2.5** The typical construction of TMF contacts with current and magnetic paths.

**AMF contacts**

Techniques to produce axial magnetic field are manifold. It could be realized by energization of windings installed behind the electrodes [1–3, 6]. Another approach is to use a ferromagnetic material to change the azimuthal self-magnetic field around the stems in axial direction [6]. The contacts are manufactured in slotted forms to minimize the lagging of the magnetic field due to eddy current flowing on contacts. Construction of an AMF contact pair is schematically demonstrated in Fig. 2.6. Magnetic field intensity, the arc current value and contact diameter determine which discharge mode could be appeared in interrupters developed by AMF technology. Multiply-arc is observed in low-current cases [6]. Melting of the contact is prevented as a result of formation of multiple arcs. At high-current the discharge is transformed to diffuse columnar arc.

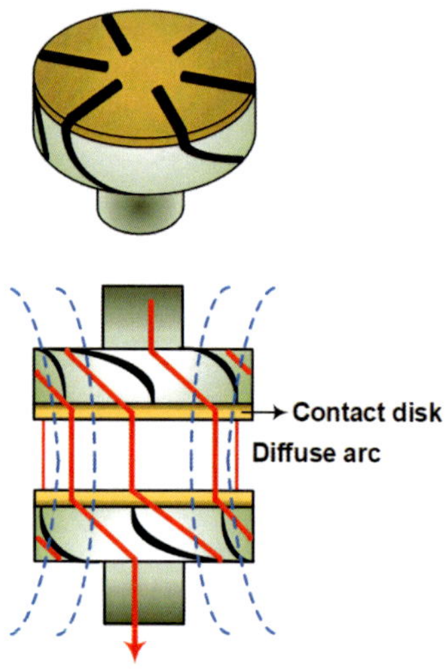


**Fig. 2.6** The typical construction of AMF contacts with current and magnetic paths.

## 2.5 Electrode materials

Contact material and surface condition of electrodes as roughness, depositions and oxides influence the breakdown voltage and interruption performance in vacuum. Contact materials, e.g. CuCr[1], Cu, AgWC, etc. are subjected to comprehensive investigations [1, 31, 32]. Due to the superior properties of CuCr, it is privileged and divers composition ratios of CuCr (e.g. CuCr7525, CuCr50, Cu, etc.) are studied with respect to the interruption performance of vacuum switches and to threshold currents of high-current anode modes [1, 13, 32, 33].

Incorporation of Cr in Cu electrode enforces mechanical integrity of the contact surface through which the electrodes could be protected against erosion

[1] It should be noted that the electrode materials are alloys and not chemical compounds. Although a notation like Cu-Cr would be preferable from the chemical point of view, CuCr has been stablished and widely used in science and technical description.

and consequently the dielectric withstand strength of the switch can be enhanced. This underlines the importance of addition of Cr. Nevertheless, modification of the electrodes with Cr deteriorates interruption performance of vacuum switches since addition of Cr can decrease substantially the thermal conductivity of CuCr admixture. Thermal conductivity of the CuCr composition is reduced from 40 to 50% by increase of weight fraction from 25 to 50 wt.% [32] .

Increase in Cu portion enhances the arc energy and consequently increases melting of the electrodes; it reduces, however, the melting duration. It should be noted that the released particles in vacuum gap could have more influence on the breakdown voltage compared to the melting duration. An optimum content ratio of Cu and Cr as well as the best strategy of production are still subject to debate.

Arcing performance of vacuum interrupter by modification of the electrode with W, C, Te, Bi, etc. are also investigated, e.g. to decrease electrode erosion or as a protection against contact welding. However, the electrodes modified by aforementioned additives have not still practically been employed in vacuum interrupters [1].

## 2.6 High voltage application

There is a great interest to develop interrupters with vacuum technology for high-voltage levels. In this regard, the not-satisfying dielectric property of the switching gap is still the crucial issue. This is addressed by two technical approaches. Enlargement of the gap distance is one approach. The other approach is based on the incorporation of two or multiple gaps in series in which each realizes as a sub-gap the required gap distance. However, it should be mentioned that the breakdown voltage does not have linear behavior over the gap length and can be estimated with the following equation [28]:

$$U_b = A_b d_b^{\theta} \tag{2.1}$$

where $U_b$ denotes the breakdown voltage, $A_b$ is a constant, and $\theta$ is a constant less than 1.

In gas filled switches like SF6 interrupters, breakdown voltage is determined by volume effect and is linearly dependent on the gap distance, i.e. regarding a critical field strength higher operation voltages can be achieved by higher electrode distances. The vacuum breakdown, however, is influenced by multiple factors as volume and surface effect due to the electrical field enhancement. In modern vacuum circuit breakers, the volume effects are reduced due to the ultrahigh vacuum conditions. With increasing voltage the surface conditions, i.e. roughness and boundaries, become the dominant parameters that are capable to enhance the field intensity by orders of magnitude and thus limit the breakdown voltage considerably [28].

Breakdown voltage of some commonly used insulation media is compared in Fig. 2.7. A linear dependency could be seen between breakdown voltage and gap length for media like SF6 and air. The breakdown voltage in vacuum gaps has the steepest increase and reaches the highest values for voltages up to 150 kV.

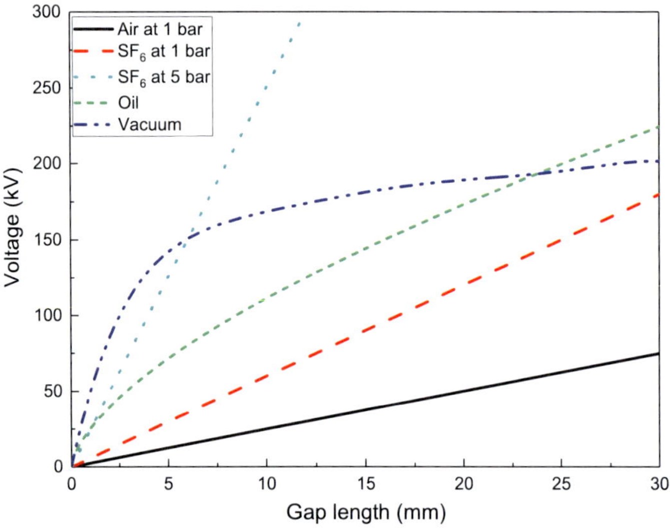


**Fig. 2.7** Breakdown voltage of different insulation media as a function of gap length [28].

It has a superior breakdown withstand strength even compared to the SF6-filled gap at 5 bar up to the gap length of about 6 mm. However, it is not linear over the gap length and limited below 200 kV. Additionally, much larger gap lengths should be provided for vacuum interrupters with increment of the voltage. Nevertheless, a larger gap would result in higher stress on the bellow, higher energy mechanism and a much bigger construction for interrupter. Furthermore, reduction of the axial magnetic field could be the outcome of the increment of the gap length.

Single gap vacuum interrupters are commonly employed for the voltages up to 72.5 kV though first commercial products for the 145 kV level are introduced recently. To use the vacuum interrupters for higher voltage levels, specifically for 145 kV and above, multiple vacuum interrupters could be integrated in series as an economical approach. A research is already launched aiming at design and test of vacuum interrupters for 145 kV and 245 kV using multiple gaps [28].

The series-connected switches allow to apply the vacuum interrupter for higher voltage stresses. However, the voltage is not distributed equally across the units because of the stray capacitance. This can result in non-simultaneous opening time of the interrupters [28]. The non-uniform voltage distribution could bring about an internal breakdown or flashover. This problem could be mitigated by incorporation of grading capacitors for uniform voltage distribution. Besides, vacuum interrupters could be over designed to stand the harsh stresses. To summarize, there has been considerable research including lab-scale prototype recently, there seems to be no actual research towards commercial multi-stage vacuum interrupters.

## 2.7 Different discharge modes in vacuum

Apart from low-current diffuse mode, three high-current discharge modes are identified for the vacuum arc: footpoint, anode spot, and intense modes [7–12, 30, 34]. Meanwhile, the anode plays an active role; the contact surface undergoes an overheating and consequently melting of anode surface occurs. The transition from the low-current mode to high-current modes can be recognize though detection of optical, thermal, and electrical changes at the anode, the cathode, and in the inter-electrode gap [8, 12]. Shown is in the Fig. 2.8 the pictures of abovementioned discharge modes at different instants of time for CuCr7525 electrodes with a diameter of 10 mm.

**Diffuse mode**

In the case of the diffuse mode, anode is passive and scavenges those particles released from cathode and those occupied the inter-electrode area in the form of plasma. Meanwhile, the cathode spots can be observed with very small luminous plasma areas that fast move on the surface of the cathode. In fact, the spots can be seen as an arc-constricted region that comprises also the cathode body and dense plasma generation area where the current continuity is supported [9]. The cathode spot can disappear and again appear with a typical spot lifetime in the range of nanoseconds. In diffuse mode, the arc voltage is relatively low with low noise level [8]. No relevant erosion is expected for the anode. Since the anode attracts materials evaporated from the cathode, it gains material in the diffuse arc mode [9]. Formation of the diffuse mode is depicted in Fig. 2.8a, where the current is about 1.5 kA.

**Footpoint mode**

The footpoint mode is an intermediate high-current anode mode where the anode starts to take an active role in the discharge. In the footpoint mode the inter-electrode gap is mostly filled with a fairly bright diffuse glow but brighter compared to the diffuse mode [9]. Single spot or multiple luminous spots can be observed on the anode in footpoint mode, however, it is not the case for the diffuse mode with no light emission from the anode [9, 10]. These bright spots are known as footpoints. A footpoint is characterized as being a small luminous spot due to the local anode melting and with the migration of the anode material to the arc channel.

This mode is pictured in Fig. 2.8b with applied current about 2.6 kA. In the meantime, considerable increase of mean discharge voltage occurs in footpoint mode, which is influenced by contact material. This goes hand in hand with rise of the arc voltage fluctuation. [9, 12].

**Anode spot mode**

The anode spot mode is classified in high-current modes with high anode activity. In the anode spot mode sometimes an arc column can be distinguished in the gap area. One large or several small very bright spots appear on the anode which could merge to a single anode spot if the discharge current or time would be increased [9]. Temperature of the anode spot exceeds the melting point and may reach the value near the atmospheric boiling point [12]. Unlike the footpoint mode, during an anode spot a large amount of particles is released to the inter-electrode gap due to the higher temperature (and thus vapor pressure) and the larger involved electrode surface [9–11].

In the anode spot mode the arc voltage is normally low and without fluctuation but may reach a relatively high value with superimposed fluctuations [9]. In this case the arc discharge voltage is higher than in the diffuse arc mode. Fig. 2.8c shows the appearance of anode spot by applied current of 3.2 kA.

**Intense mode**

The intense arc mode belongs to the high-current modes with higher activity of the anode compared to anode spot mode. In the intense mode, very luminous areas surround both the anode and the cathode and can be also observed in the inter-electrode gap [9–11]. The mode takes place at smaller gap distances compared to the anode spot. Hence, jets emitted from anode and cathode cannot be differentiated. During this mode massive erosion of both anode and cathode is detected; when the temperature of electrodes approaches the boiling point at atmospheric pressure [12]. The discharge voltage remains low and without considerable voltage oscillation; it is, however, bigger compared to that in the diffuse mode. A shot of the intense mode is shown in Fig. 2.8d where the applied current is about 4.5 kA.

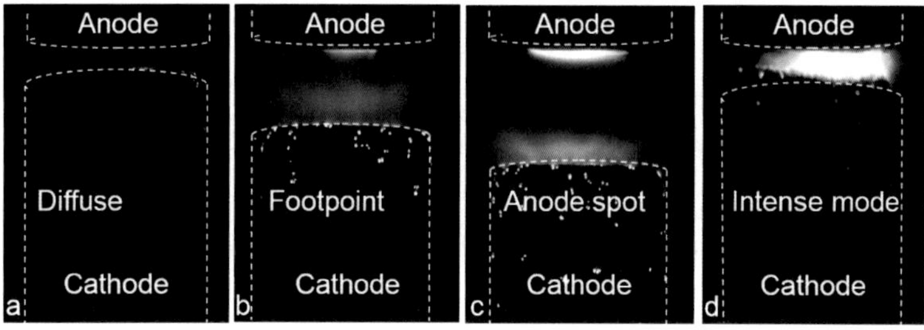


**Fig. 2.8** The various discharge modes in the vacuum arc for different shots for AC 50 Hz and CuCr7525 with diameter of 10 mm.

## 2.8 Influence of magnetic field on high-current anode modes

Formation of high-current anode modes can be affected by magnetic fields (TMF and AMF). In this regard, the direction of the magnetic field plays the most important role [35]. An axial magnetic field leads to reduction of the losses of charged particles to the sides of the inter-electrode gap. By applying a weak axial magnetic field the overall arc voltage is decreased due to reduced plasma loss in the gap and correspondingly greater plasma flow to the anode. The reduction of plasma losses produced by axial magnetic field can increase the threshold current of the formation of high-current anode modes [2]. However, a strong axial magnetic field will cause constriction of the current and leads to increase the arc voltage. Under this condition the ratio of column surface area to column volume becomes larger which leads to the increase of the loss from the arc column. This loss should be compensated by increase of the arc voltage [2].

In a transverse magnetic field the current flows through the contacts resulting in a magnetic field component perpendicular to the arc. The most common application of transverse magnetic field is to move columnar arc [1, 2]. Therefore, the high-current anode modes does not stay in one point of the anode, which can reduce the anode erosion by preventing the localized melting of the anode surface and thus, increase the threshold current of the formation of high-current anode modes.

## 2.9 State of the art in the research of high-current anode phenomena

Transitions from a diffuse low-current mode to high-current modes have been investigated by electrical measurements and cinematography [13, 14]. Discharge modes can be detected by some changes in optical, thermal, and electrical characteristics of the arc, such as the light emitted by the region near the anode, arc voltage characteristics, and the anode surface temperature [9, 12–14, 30, 34, 36–38].

The appearance of divers discharge modes in vacuum interrupters can be presented in an existence diagram (gap length versus current) [8, 9, 11, 12]. In the existence diagram, the area in which different high-current anode modes appear are roughly separated by boundary lines. These boundaries can be altered by contact material, frequency of interrupting current, and electrode geometries, etc. [9, 14, 39, 40].

The frequency of the interrupting current affects the threshold current of high-current anode phenomena formation. By increasing the frequency, the threshold current will be increased. In contrast, in long pulses or DC waveforms, the threshold current will be decreased [9].

Based on the electrode displacement curve (time dependent of gap length versus current), the appearance of high-current modes can be prevented by a proper selection of system parameters. In fact, by changing the opening speed, the arc can remain in diffuse mode or footpoint mode in which the erosion rate is still much less than in the anode spot mode or intense mode [39].

The electrode material can also alter the threshold currents of the appearance of high-current anode phenomena. Moreover, by increasing the electrode diameter, the threshold current as well as the electrode distance, at which the high-current anode phenomena appear, increases considerably [33, 40].

Regarding the impact of anode surface temperature on the generation of metal vapor, the anode surface temperature is suitable to allow conclusions on the dielectric recovery strength after current zero by the aid of metal vapor evaporation estimations [15]. The anode surface temperature during and after zero-crossing shows that the decay of the anode surface temperature after current zero in vacuum arcs can be classified as two modes [24–26]. First mode: The anode surface temperature decays exponentially after current zero; and second mode: The temperature drops from the initial temperature at current zero to 1370 K, which is close to the melting point of Cu, then the temperature remains constant for a period of time, and the temperature exponentially drops from 1370 K to the room temperature [24, 25]. In fact, the first decay mode describes the anode surface decay process after extinguishing a low-current arc, and the second mode describes the process after extinguishing a high-current arc where parts of the electrode are melted.

A last anode effect shortly before current zero is anode plume which has been introduced recently by the group of Batrakov [41–43]. By anode plume, a shell-like boundary appears in front of the anode in case of CuCr electrodes [41]. Emission from this shell is basically identified to be originating from neutrals. A plume is covered by a halo that is formed by spectral lines from ionic particles. However, up to now the anode plume was only investigated by framing camera with high temporal resolution but very limited number of frames using interference filters for observation of atomic and single charge ionic line emission. No spectroscopic analysis has been applied so far [43].

To understand the physics of high-current anode phenomena, some optical diagnostics including emission spectroscopy could be conducted to investigate different atomic and ionic states near the cathode, the anode, and in the inter-electrode gap [17–21, 44]. By transition from low-current modes to high-current modes or even between different high-current modes, the intensity of atomic and ionic species will be changed at the anode, the cathode, and in the inter-electrode gap [17].

The temporal evolution of a hot-anode vacuum arc discharge (HAVA) was investigated in case of pulsed DC with current of 175-340 A and duration of 150-200 s. Copper electrodes were used for experiments. Spectral lines of Cu I, Cu II and Cu III have been investigated as a function of time near the cathode, near the anode and in the middle of the electrode gap [17].

Different emission lines were measured for a vacuum arc at diffuse, anode spot, and intense modes during a single pulse in [18]. However, higher ionization states (Cu III) that may be important to the arc dynamics could not be readily observed with the experimental setup. Al and Cu electrodes were used for cathode and anode respectively and AC pulses with a duration of 8.5 ms were applied. At the transition to anode spot mode, Cu emission is increased and Al emission is decreased. Nevertheless, both Al and Cu emission are increased at the transition to the intense arc mode. Spatial profiles of the excited-state density and Boltzmann distribution temperatures were used at different positions in the gap [18]. It should be noticed that LTE might be not applicable in such a high-current mode.

A diffuse mode in vacuum arcs with CuCr electrodes has been investigated by analyzing spectra of six species Cu I, Cu II, Cu III, Cr I, Cr II, and Cr III in the wavelength range 350–810 nm [21]. The axial intensity distributions were strongly dependent on the ionization stage of radiating species. Emission distributions of Cr II and Cu II can be distinguished as well as the distributions of Cr III and Cu III [21]. The axial distribution along the gap was determined for different atomic and ionic Cu and Cr lines. However, this investigation was applied only to a diffuse arc mode. The arc behavior for CuCr electrodes was spatially resolved by optical emission spectroscopy along the gap axis. Different intensity distributions of atomic and ionic lines were found along the gap [21].

The atomic line emission of both copper and chromium is much stronger near the electrodes than in the gap. However, the intensity of ionic lines is higher in the gap [20]. Spectra simulations indicate that the plasma layer in front of the anode is not in LTE and best agreement is found for a two-temperature plasma with a total pressure of 0.03 bar as obtained from the Stark broadening of a copper atomic line, an electron temperature of 15 000 K and a ratio of electron to gas temperature of 1.3 [20]. However, there is still a lack of investigation regarding optical diagnostics especially during high-current anode modes.

## 2.10 Numerical models

### Physical models

Numerical modeling of vacuum arc discharges plays an important role in understanding of physical phenomena taking place in the plasma. It gives the information, which cannot be acquired in experiment, like e.g. plasma composition, properties of near-electrode regions, temperature distribution in the plasma. Different kind of models have been used in the past for consideration of cathode region properties and column properties. Complex 3D models for vacuum arcs affected by magnetic field are available [45–49]. However, the anode phenomena are generally not included in such models. Recently the model for anode "deformation" is suggested under the action of plasma pressure. The anode sheath is considered to be homogeneous with rough approximation of energy flux. The model which explains the high-current anode phenomena in general, and transition between different anode modes is not available. Beside pure scientific interest, the clarification of appearance conditions for different anode modes has also well pronounced practical aspects. Spatial and temporal evolution of the species densities emitted from the anode close to current zero has direct impact on interruption failure in vacuum circuit breakers.

### Electrical arc models

In contrast to the more complicated models, there are some simple electrical arc models in which the discharge process including different physical processes are described only by a first-order differential equation (that may not completely describe the arc from the physical point of view). The simplified electrical arc models (denoted further in text as conventional models) are based on the Cassie and Mayr arc models [50, 51]. The Mayr arc model is typically used for investigation of the dynamic regime of the electric arc in the low-current ranges, i.e. around current zero [50] and is presented in (2.2), whereas the Cassie arc model [51] which is typically applied in the high-current ranges is presented in (2.3).

$$\frac{1}{g_m}\frac{dg_m}{dt} = \frac{1}{\tau_m}\left(\frac{ui}{P_0} - 1\right) \tag{2.2}$$

$$\frac{1}{g_c}\frac{dg_c}{dt} = \frac{1}{\tau_c}\left(\frac{u^2}{u_0^2} - 1\right) \tag{2.3}$$

where $u$ and $i$ are voltage and current of the arc respectively, $g_m$ and $g_c$ determine the conductance of the Mayr and Cassie models, $\tau_m$ and $\tau_c$ represent time constants of the Mayr and Cassie of the electric arc, $P_0$ stands for the dissipated power at current zero, $u_0$ is the average arc voltage. In principle, $u_0$ determines the average value of arc voltage during high-current regime, where the arc time constant and the dissipated power in the Mayr model have impact on arc voltage behavior during current zero-crossing such as extinguishing voltage, TRV and RRRV.

The Habedank arc model is a combination of the Mayr and the Cassie arc model to cover both the current zero and high-current areas of the arc [52]. Some hybrid models based on the Habedank arc model use weight functions for switching between the Cassie and the Mayr model, and these weight functions can also increase the number of estimated parameters [53–56]. In [57] some weight functions based on arc current and desired switching current are represented.

The Schavemaker arc model presented in (2.4) is almost similar to the Habedank model. Here $\tau$ is arc time constant, $U_{arc}$ is average arc voltage, $P_1$ is cooling constant. In the low-current phase, the term $P_0$ remains in the equation; so, it is similar to Mayr model. However, in high-current area, the term $U_{arc}|i|$ is the one that remains and it defines Cassie arc model [58].

$$\frac{1}{g_s}\frac{dg_w}{dt} = \frac{1}{\tau_s}\left(\frac{ui}{max(U_{arc}|i|, P_0 + P_1 ui)} - 1\right) \tag{2.4}$$

The Schwarz arc model presented in (2.5) is actually a Mayr-based arc model in which the dependencies of power loss and arc time constant from the arc conductance are defined by the variables $a$ and $b$ [59]:

$$\frac{1}{g_{sw}}\frac{dg_{sw}}{dt} = \frac{1}{\tau g_{sw}^a}\left(\frac{ui}{P g_{sw}^b} - 1\right) \tag{2.5}$$

where $P$ is dissipated power at current zero, $a$ and $b$ are calibration factors.

The KEMA arc model is derived from three Mayr or Cassie arc models in series and is shown in (2.6). In this model, the Mayr or the Cassie arc models can be selected by assigning different values to $\lambda_i$, where $\lambda = 1$ defines the Mayr arc model and $\lambda = 2$ points out to the Cassie arc model [59, 60]:

$$\frac{1}{g_k} = \frac{1}{g_1} + \frac{1}{g_2} + \frac{1}{g_3}, \qquad \frac{dg_i}{dt} = \frac{A_i}{\tau_i} g_i^{\lambda_i} u_i^2 - \frac{1}{\tau_i g_i} \tag{2.6}$$

where $\tau_2 = \tau_1/k_1$, $\tau_3 = \tau_2/k_2$ and $A_2 = A_3/k_3$, $k_n$ are calibration factors, and $A_n$ is a cooling constant of n-th arc.

The modified-Mayr model is more similar to the Schavemaker model and is defined in (2.7). Therefore, by altering the current values, the Mayr or the Cassie arc model will be singled out [59]:

$$\frac{1}{g_{mm}}\frac{dg_{mm}}{dt} = \frac{1}{\tau_{mm}}\left(\frac{ui}{p(P_0 + C_i|i|)} - 1\right) \tag{2.7}$$

where $C_i$ is current constant and $p_{mm}$ is a pressure of circuit breaker.

In Voronin-Sawiki's Hybrid model which is presented in (2.8), the arc geometry is considered. In this model the arc time constant and the dissipated power are proportional to the arc volume, and the arc geometry variation in time is also considered [57]:

$$\begin{aligned}\frac{1}{g_{vs}}\frac{dg_{vs}}{dt} = \frac{1_{vs}}{\tau_{vs}d(i)}&\left\{[1-\varepsilon(i)]\frac{ui}{P_V(l_{vs},d(i))} + \varepsilon(i)\frac{ui}{P_s(l_{vs},d(i))} - 1\right\}\\ -\frac{1}{l_{vs}}\frac{dl_{vs}}{dt}&\left(1 + ln\frac{4g_{vs}l_{vs}}{\pi K_g d^2(i)}\right) + \frac{2}{d(i)}\frac{dd(i)}{dt}\left(1 - ln(\frac{4g_{vs}l_{vs}}{\pi K_g d^2(i)})\right)\end{aligned} \tag{2.8}$$

where $d(i)$ and $l_{vs}$ are diameter and the length of the arc, respectively. $\varepsilon(i)$ is weight function, $P_S$ and $P_V$ are density of power dissipated through the lateral surface and volume of the arc column, respectively, and $K_g$ is coefficient of approximation of unitary conductance. This model is more complicated than the models described in (2.2) to (2.7). However, the arc geometry and its variation in time are considered in the model [61].

The arc model based on physical parameters is proposed and investigated in [62–64]. This model considers different power contributions, e.g. radiative, turbulent and power carried by axial and radial mass flows. In this model, physical parameters of the arc such as gas density, pressure, enthalpy, temperature, and gas flow velocity are taken into account [63, 64]. The time dependent total power balance of the arc is calculated as follow

$$P_{total} = P_{rad} + P_{tur} + P_{ramf} - P_{axmf} + P_{net} \tag{2.9}$$

where $P_{rad}$ is the radiative power loss, $P_{tur}$ is the turbulent power loss, $P_{axmf}$ is the power gain carried by axial mass flow, $P_{ramf}$ is the power loss carried by radial mass flow and $P_{net}$ is net power of the arc. The radiative power loss is described by the Stefan-Boltzmann law or net emission coefficient [65, 66]. The turbulent power loss is produced at the surface of the arc. They can be calculated as [63]

$$P_{rad} = 2\pi\, k_{rad} \cdot \sigma_B \cdot l_{arc} \cdot r_{arc} \cdot T_{arc}^4 \tag{2.10}$$

$$P_{tur} = k_{tur} \cdot h_{arc} \cdot \rho_{arc} \cdot l_{arc} \cdot r_{arc} \cdot v_{sound} \tag{2.11}$$

$$P_{axmf} = k_{axmf} \cdot \dot{m}_{ax} \cdot h_{arc} \tag{2.12}$$

$$P_{ramf} = k_{ramf} \cdot \dot{m}_{rad} \cdot h_{av} \tag{2.13}$$

$$P_{net} = k_{net} \cdot l_{arc}[\frac{d}{dt}(\rho_{arc} \cdot A_{arc} \cdot h_{arc})] - \dot{p}_{arc} \cdot A_{arc} \tag{2.14}$$

where $\rho_{arc}$ is the mass density in the arc, $p_{arc}$ is the arc pressure, $h_{arc}$ is the enthalpy in arc, $h_{av}$ is the temporal average enthalpy of arc, $l_{arc}$ is the arc length, $r_{arc}$ is the arc radius, $A_{arc}$ is the arc cross section, $T_{arc}$ is the arc temperature, $\dot{m}_{ax}$ is the axial mass flow, $\dot{m}_{rad}$ is the radial mass flow, $v_{sound}$ is the speed of sound in the arc, and $\sigma_B$ is the Stefan-Boltzmann constant. It should be noticed that $k_{rad}$, $k_{tur}$, $k_{axmf}$, $k_{ramf}$, and $k_{net}$ are calibration constants.

In order to evaluate the power-based model, the variables of this model must be determined. However, regarding the number of variables, the application of a parameter estimation method like a least square nonlinear method causes unexpected errors. However, some of these variables such as $A_{arc}$, $l_{arc}$ and $T_{arc}$ can be determined by diagnostics like optical diagnostics or spectroscopy. [67, 61, 68] can be used to estimate parameters of power based model.

The arc model based on the arc diameter is proposed in [69]. Usually the arc diameter is considered as an implicit parameter in the arc model but the arc diameter is current-dependent and increases when the current grows. In the model based on the arc diameter, the arc diameter is considered as direct parameter that can cover both the high and low current area, which is usually described by combination of Mayr and Cassie models.

Generally, the arc cross-section is a function of the current $(A \propto |i|)$ [69]. Considering the arc column as cylindrical, the arc diameter can be described as $d_{arc} \propto |i|^{0.5}$ assuming constant current density. Hence it is defined as:

$$d_1 = \beta |i|^{1/2} \tag{2.15}$$

where $\beta$ is constant parameter in mm and $i$ is the current. Since the dependency of the arc diameter on the current can be a function of additional effects and corresponding diameter variation, formula (2.16) is proposed [69] which requires numerical values for $\beta$ and $\alpha$:

$$d_2 = \beta |i|^{\alpha} \tag{2.16}$$

More formulas can be found in [63, 69–72]. The general form of the arc model based on the arc diameter can be defined by the following equation:

$$\frac{1}{g_d}\frac{dg_d}{dt} = \frac{1}{\tau_d}\left(\frac{ui}{U_{arc}|i| \cdot (a \times d_k^c)} - 1\right) \tag{2.17}$$

where $a_d$ and $c_d$ are additional calibration factors determined from the experiment. By selecting the proper values for the calibration factors, the arc voltage in high current range and at the zero-crossing can be calculated.

# Chapter 3

# Methodology and theory

## 3.1 Experimental setup

In this chapter, the experimental setup including high-current generators, electric measurement systems and vacuum chamber is explained. Besides, modus operandi of the employed optical diagnostic systems including emission spectroscopy, absorption spectroscopy, and high-speed cinematography is described. Moreover, the basic theories which are adopted to analyze the optical measurements are presented.

### 3.1.1 High-current generators

In this work, alternating current (AC) pulses of 50, 100, 180, and 260 Hz beside direct current (DC) pulses over 5 and 10 ms are submitted to the implemented electrode arrangement to study the high-current anode modes. For this purpose, two groups of both capacitors and coils are connected to different discharge circuits.

**AC pulses**

An LC circuit can be used to generate damped AC. The following equation is applicable for the LC circuit

$$L\frac{di}{dt} + Ri + d\frac{q}{C} = 0 \tag{3.1}$$

considering $i = dq/dt$, equation (3.1) can be written as

$$\frac{d^2q}{dt^2} + \frac{R}{L}\frac{dq}{dt} + \frac{q}{LC} = 0 \tag{3.2}$$

By solving (3.2) the electric load can be obtained as follow

$$q = q_{max} \cdot e^{-\frac{R}{2L}t} \cdot cos(\omega t) \tag{3.3}$$

where $q_{max} = C \cdot V_c(0)$, $C$ is the capacitance of capacitor bank, $L$ is inductive part of coil, $R$ is the resistive part of coil and connectors, $V_c(0)$ is the charging voltage of capacitors, and $\omega$ is angular frequency which can be determined from

$$\omega = \sqrt{1/LC - (R/2L)^2} \tag{3.4}$$

Regarding (3.3) and $i = dq/dt$, the current can be written as

$$i = -\frac{q_{max} \cdot e^{-Rt/2L} \cdot [2L \cdot \omega \cdot sin(\omega t) + R \cdot cos(\omega t)]}{2L} \tag{3.5}$$

The maximum current for LC circuit can be delivered as

$$i_{max} = q_{max} \cdot \omega \tag{3.6}$$

The schematic representation of the high-current generator for AC pulses is shown in Fig. 3.1. The capacitance of each capacitor is about 87.5 $\mu$F. Each 6 parallel capacitors (525 $\mu$F) build a block and each block is connected in series with one coil.

As presented in (3.4), by changing the values of $C_{in}$ and $L_{DAC}$, different frequencies can be obtained. Here, this can be realized by changing the number of parallel capacitors and selecting different coils. Notice, for AC 50 Hz an additional coil $L_{zus}$ should be incorporated in series with the LC circuit (see Fig. 3.1).

Typical waveforms of damped AC pulses generated in INP's high-current laboratory are depicted in Fig. 3.2. Depending on the behavior or properties of the test object, the damping can be increased. Notice that currents with other frequencies, e.g. 200 Hz, 16.7 Hz, 25 Hz, etc. can be also generated in this laboratory.

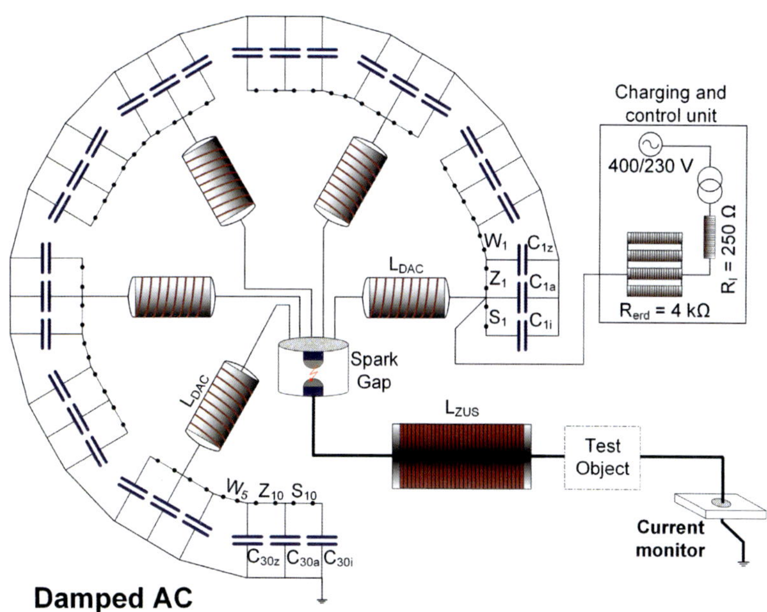


**Fig. 3.1** The schematic representation of the high-current generator for AC pulses.

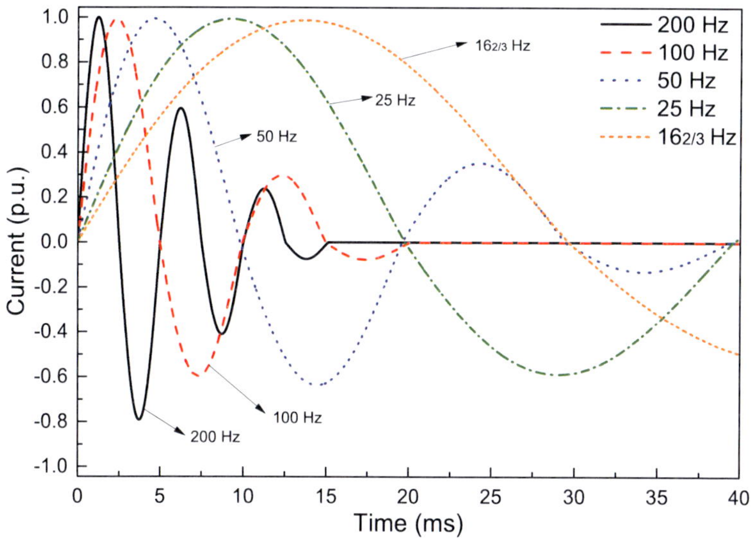


**Fig. 3.2** Typical damped AC pulses generated in the high-current laboratory.

The values of $C_{in}$, $L_{DAC}$, $L_{zus}$, and $n$ introduce different frequencies which are used in this work and provided in Table 3.1.

**Table 3.1** Parameters of high-current generator for different AC waveforms which are used in this work.

| Waveform | $C_{1..n}$ [$\mu$F] | $L_{DAC}$ [$\mu$H] | $L_{ZUS}$ [$\mu$H] | Parallel stages [n] |
|---|---|---|---|---|
| AC 50 Hz | 525 | 4700 | 2730 | 5 |
| AC 100 Hz | 525 | 4700 | — | 5 |
| AC 180 Hz | 262.5 | 4700 | — | 3 |
| AC 260 Hz | 87.5 | 4700 | — | 3 |

## DC pulses

A rectangular pulse can be produced by using of series connected LC elements forming an n-stage ladder network and discharging into the terminating resistor *R*. The values of terminating resistor and the approximate duration of the pulse can be calculated by

$$R = \sqrt{L/C} \tag{3.7}$$

$$T_{DC} \approx 2\frac{n}{n-1}\sqrt{LC} \tag{3.8}$$

with $L = nL_i$ and $C = nC_i$ , where $L_i$ and $C_i$ are the inductance and the capacitance of each stage, respectively. Hereof, an asymmetrical combination of the ladder network is however more advantageous in order to obtain, as far as possible, a rectangular current pulse without large overshoots or undershoots during the pulse. Moreover, the value of the terminating resistor is affected by the resistance of a nonlinear resistance of the arc.

In the high-current laboratory, a high-current generator for DC pulses is constructed with LC elements having constant capacity of 262.5 $\mu$F and stepwise-decreasing inductivities in 10 stages (Fig. 3.3).

Typical DC pulses generated in INP's high-current laboratory are plotted in Fig. 3.4.

Obviously, the shape of the pulses presented in Fig. 3.4 is not an ideal pulsed DC (perfectly rectangular). The rise time is about 1 ms and fall time is about 2 ms. However, the pulse is named as DC 5 ms or DC 10 ms what is corresponding to the DC part of the pulse. Therefore, the term "pulsed DC" conveys this fact that the discharge current remains almost constant during most of the discharge process.

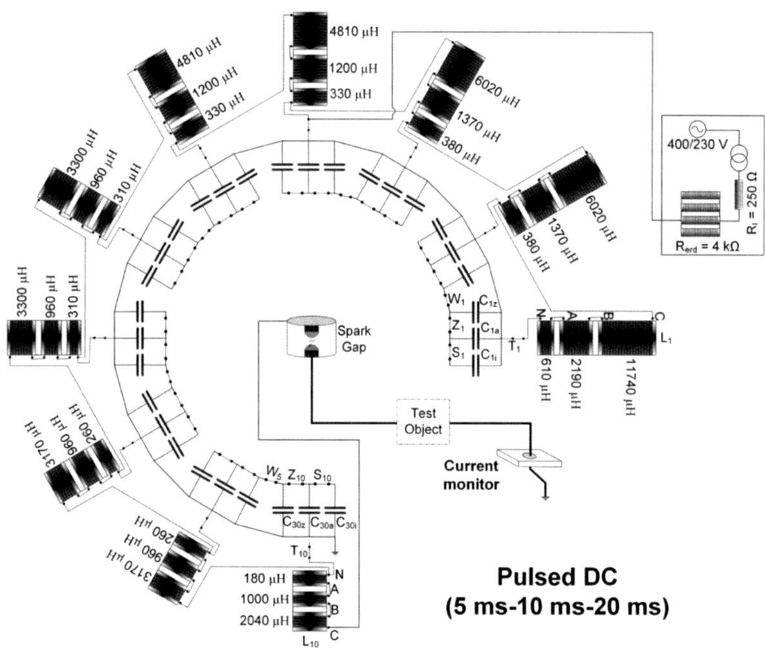


**Fig. 3.3** The schematic representation of the high-current generator for DC pulses.

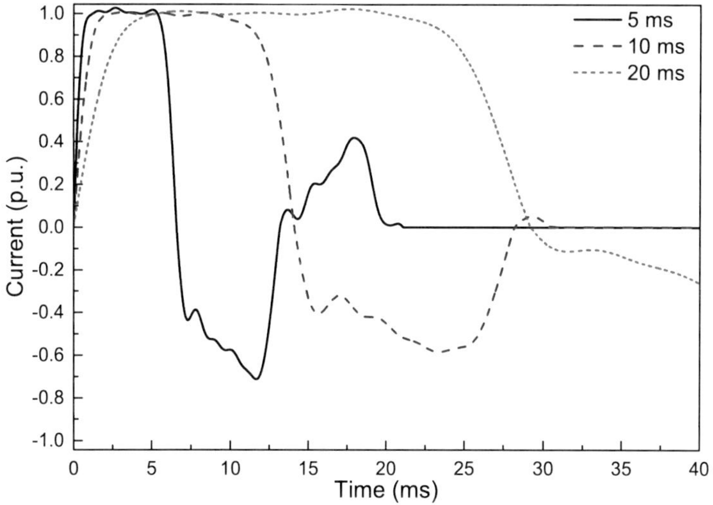


**Fig. 3.4** Typical DC pulses generated in the high-current laboratory.

The corresponding inductances to generate pulsed DC over 5 ms, 10 ms, and 20 ms are delivered in Table 3.2.

**Table 3.2** The corresponding inductances to generate pulsed DC over 5 ms, 10 ms, and 20 ms.

| Element | Inductance [$\mu$H] | | |
|---|---|---|---|
| | 5 ms | 10 ms | 20 ms |
| $L_1$ | 610 | 2800 | 14540 |
| $L_2$ | 380 | 1750 | 7770 |
| $L_3$ | 380 | 1750 | 7770 |
| $L_4$ | 330 | 1530 | 6340 |
| $L_5$ | 330 | 1530 | 6340 |
| $L_6$ | 310 | 1270 | 4570 |
| $L_7$ | 310 | 1270 | 4570 |
| $L_8$ | 260 | 1220 | 4390 |
| $L_9$ | 260 | 1220 | 4390 |
| $L_{10}$ | 180 | 1180 | 3220 |

**Operation principle**

The first step to generate high-current pulses is to charge the capacitors. The capacitors can be charged up to 18 kV. To produce different current levels based on the required frequency, different charging voltages are applied. To discharge the capacitors, the spark gap in ambient air with electrodes made out of WCu is connected to the circuit. The spark gap is initiated by a Marx generator. The trigger mechanism is controlled by an optical delay generator. Notice, the distance between the electrodes of the spark gap must be modified based on the charging voltage, which can be carried out by electric motor. The lower part of the spark gap is connected to the test object (vacuum chamber) through a high-voltage cable. The triggering of spark gap, pneumatic movement of vacuum chamber electrode and optical diagnostics devices are also governed by the optical delay generator and optical cables. The signals including current and voltage (see 3.12 and 3.13) are registered by a transient recorder (HBM Gen7) and analyzed using corresponding software. Since electrical signal are converted to optical and transferred by optical cables to a control room before being analyzed, no electric connection between operation and control room is needed.

## 3.1.2 Current measurement

Capturing the amplitude and waveform of the current is essential in many applications. The most common approach is integrating a resistor into the circuit and measuring the voltage drop associated to the induced magnetic field. Another practiced method is using a magnetic coupling which includes direct current probes based on the flux-gate and Hall-effect principles such as air-core Rogowski coils or transformers with high permeability.

In this work, a Pearson current monitor is used. Current monitors differ from conventional current transformers because they are internally terminated and deliver an output voltage proportional to the current. Moreover, they have only one winding, as primary winding, that surrounds the conductor. In contrast to the simple resistor, Pearson current monitors are suitable for cases in which ground-loop noise, lack of high voltage isolation, voltage drop and power dissipation are of concerns. The schematic of current monitor consisting of magnetic core, the secondary winding, the resistive termination and the electromagnetic shield is sketched in Fig. 3.5.

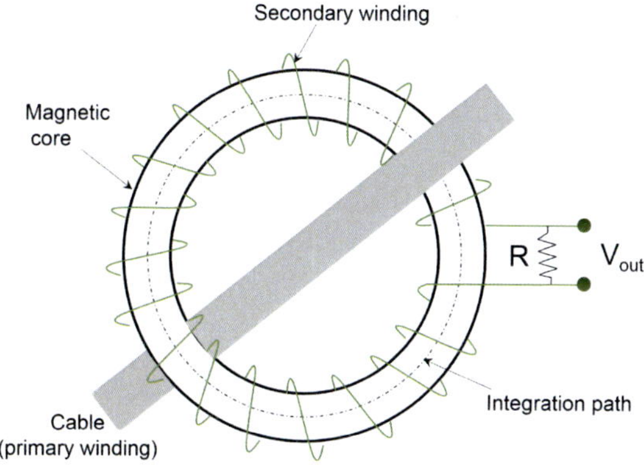


**Fig. 3.5** The schematic of current transformer terminated with a resistor.

Considering Fig. 3.5, it is assumed that the toroidal high permeability magnetic core has $n$ turns. The primary circuit is the conductor going through the magnetic core and carrying the high-current. The magnetic field of this conductor is nearly radially symmetric. The cross-section area of the core is negligible compared to the radius if the toroid. Based on the ampere's law

$$H = (i_p - ni_s)/l_c \tag{3.9}$$

where $H$ is magnetic field, $i_p$ and $i_s$ are the primary and secondary current, and $l_c$ is the mean path length within the core. If $A_c$ stands for the effective cross-section area of the core and permeability $\mu$ is substituted with $B/H$, then equation (3.9) can be rewritten as

$$\phi = BA = \mu HA = \mu A_c(i_p - ni_s)/l_c. \tag{3.10}$$

Now, Faraday's law

$$V = n\frac{d\phi}{dt} = i_s R \tag{3.11}$$

is applied considering the number $n$ of the coil's windings. Introducing the right part of (3.10) into (3.11) yields

$$i_s R = n\frac{d}{dt}[\mu A_c(i_p - ni_s)/l_c] = \left(n^2\mu\frac{A_c}{l_c}\right)\cdot\frac{d}{dt}\left[\frac{i_p}{n} - i_s\right]. \tag{3.12}$$

Considering $L = n^2\mu A_c/l_c$ then (3.12) can be written as $di_s/dt + (R/L)i_s = (1/n)di_p/dt$. By applying the Laplace transform and some simplifications, (3.12) can be formulated as

$$I_s(s) = \left[\frac{s}{n}\right]\left[\frac{I_p(s)}{s+a}\right] \tag{3.13}$$

where $a = R/L$. The transfer function $s/n(s+a)$ is a simple high pass filter with a cutoff frequency of $a/2\pi$. Current is measured by the Pearson current monitor model 1423 with technical specifications listed in Table 3.3.

**Table 3.3** Technical specifications of Pearson current monitor model 1423.

| **Specification** | **Values** | **Units** |
|---|---|---|
| Sensitivity | 0.001 | V/A |
| Output resistance | 50 | Ohm |
| Maximum peak current | 500 | kA |
| Maximum RMS current | 2500 | A |
| Current time product | 75 | A.s |
| Low frequency 3 dB cut-off | 1.0 | Hz |
| High frequency 3 dB cut-off | 1.2 | MHz |

### 3.1.3 Voltage measurement

To measure the voltage of the arc and the spark gap, two devices are used: a capacitive-resistive voltage divider (High-Volt Dresden) and a voltage probe (Tektronix 6015A).

The capacitive-resistive voltage divider is normally used for measuring high-voltage due to its sensitivity. The single line diagram of the voltage divider is illustrated in Fig. 3.6. The specifications of different resistors and capacitors together with corresponding values are included in Table 3.4.

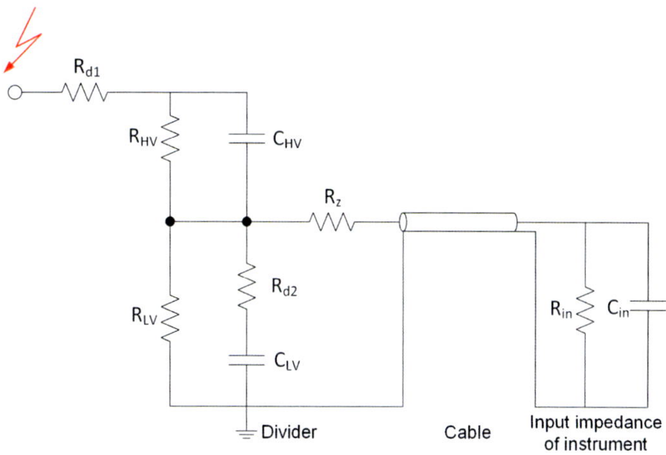


**Fig. 3.6** The single line diagram of capacitive-resistive voltage divider.

**Table 3.4** The technical specifications of capacitive - resistive voltage divider.

| **Specification** | **Parameters** | **Units** |
|---|---|---|
| External damping resistor | $R_{d1}$ | 470 Ω |
| Built-in damping resistor of the low voltage capacitor | $R_{d2}$ | 4.3 Ω |
| High voltage capacitance | $C_{HV}$ | 0.10 nF |
| Capacitance of low voltage part | $C_{LV}$ | 106.5 nF |
| Capacitance of measuring cable | $C_c$ | 2.5 nF |
| High voltage resistance | $R_{HV}$ | 50.07 MΩ |
| Low voltage resistance | $R_{LV}$ | 0.82 MΩ |
| Termination resistor at the input of the cable | $R_z$ | 50 Ω |
| Input resistance of the instrument | $R_{in}$ | 1 MΩ |
| Input capacitance of the instrument | $C_{in}$ | <50 pF |
| Resulting low voltage resistance | $R_{LV\,res}$ | 0.45 MΩ |

The transfer function of capacitive and resistive parts of the voltage divider is presented in (3.14) and (3.15), respectively. Based on the values provided in Table 3.4, the capacitive and resistive factors are obtained 111.2 and 112.0, respectively. These factors are used as setting of HBM recorder to get a corresponding voltage amplitude.

$$F_{DC} = \frac{C_{HV} + C_{LV} + C_c}{C_{HV}} \tag{3.14}$$

$$F_{DR} = \frac{R_{LVres} + R_{HV}}{R_{LVres}} \tag{3.15}$$

### 3.1.4 Vacuum chamber

The vacuum chamber consists of an ultrahigh vacuum chamber (UHV) and a pneumatic actuator for the electrode movement (cf. Fig. A2 in appendix A). The box-shaped vacuum chamber part is equipped in the upper with divers ports for electrical feed through and mounting adapters for electrodes (the upper electrode is isolated for connection to high voltage), viewports, manipulator for probes and diagnostics, isolated holder for shielding, pressure gauge and pumping. The chamber is evacuated by a small turbo-molecular pump (60 l/s) in combination with a 4-stage diaphragm pump (50 l/min). During experiments these mechanical pumps are separated from the vacuum by a gate valve and switched off to protect them from both eroded electrode material and mechanical shock caused by the pneumatic actuator [20].

An ionic pump with a pumping speed of 150 l/s reduces the pressure inside of the vacuum chamber to about $2 \times 10^{-8}$ mbar. Further improvement of the vacuum could be achieved by baking out. The ionic pump can also be switched off or protected by means of a gate valve for few minutes, e.g. when applying discharges with higher currents producing high amount of metal vapor. Two sight-glasses with a free diameter of about 150 mm allow an optical observation of the complete electrode area from two sides. These windows are widely protected from droplet and plasma deposition by exchangeable glass plates. The vacuum chamber is set on the potential of the bottom electrode and is isolated from the grounded chassis containing the actuator, pumping system, and controls. Measurement of the discharge current can be performed by means of a shunt measurement circuit or Pearson current monitor.

A mechanical-pneumatic driver system is designed and implemented for separation of the electrodes. The upper electrode, which is connected to the power supply, is fixed, whereas the lower electrode is mounted with a flexible bellow inside the UHV chamber and can be moved either up or down by the isolated pneumatic actuator. In the experiments, the top, fixed electrode serves as anode, while the lower moveable one as the cathode. A stock volume of the pneumatic actuator is filled by pressurized air of several bars and released by means of a triggered electromagnetic switch. Constant opening velocities are obtained that can be chosen between 1-4 m/s depending on the pressure in the actuator. The maximum electrode distance is about 20 mm. The delay between electrode separation and current can be adjusted freely with a total jitter below 100 $\mu$s.

In this work a butt electrode made out of CuCr7525 with a diameter of 10 mm is used to carry a similar current density existed in real circuit breaker with TMF or AMF and higher electrode diameter.

## 3.2 Optical diagnostic methods

Different optical diagnostic techniques, specifically optical emission spectroscopy (OES), high-speed cinematography, video spectroscopy, and broadband absorption spectroscopy are employed in this work. In the following, the experimental setup is presented.

### 3.2.1 Optical emission spectroscopy

The basic function of a spectrometer is taking the light and then converting it to spectral components as a function of wavelength. First, the light is directed to a narrow aperture (entrance slit) with widths from 5 to 200 $\mu$m. Then the light is collimated by a concave mirror (collimating mirror) and redirected onto a grating. Accordingly, the grating disperses the spectral components of the light at slightly varying angles due to optical diffraction. The grating determines the wavelength range and the partial optical resolution. The diffracted light is focused by a second concave mirror (focusing mirror) and imaged thereafter onto the detector. The detector converts the photons into electrons and then the corresponding current is digitized and read out by a computer. Finally, using the software the data will be prepared as a function of wavelength based on the interpolating the number of pixels in the detector and the linear dispersion of the grating.

In this work, in most cases imaging spectrometers with focal length of 500 mm or 750 mm are utilized. The spectral resolution depends mainly on the (selectable) grating (number of line per mm), width of entrance slit, and detector with its pixel size. Typically having different gratings including 150, 1200, 1800, and 2400 line per mm are applied. The spectrograph with different parts, i.e. entrance slit, mirrors, grating, ICCD camera is rendered in Fig.3.7 (cf. Fig. A3 of appendix A).

The spectrograph can be equipped with different cameras, e.g. a charge-coupled device (CCD), an intensified CCD (ICCD), a complementary metal-oxide semiconductor (CMOS), and electron multiplying CCD (EMCCD). A CCD is an array of photodiodes (light-sensitive pixels) which are electrically biased. They can generate and store electrons (electric charges) when they are exposed to the light. The amount of charge trapped under each pixel relates directly to the number of photons illuminating the pixel. By changing the electrical bias of an adjacent pixel, this charge can migrate to the next pixel as in a shift register. When the charge travels out of the sensor, it is converted into a voltage by a charge amplifier and is then digitized into a numerical values by an analog-digital converter. Thereby, high dynamic and accuracy can be obtained (up to 16 bit). This action is performed for each pixel to create an electronic image. Bearing this in mind, the electronics inside the camera controls the readout process.

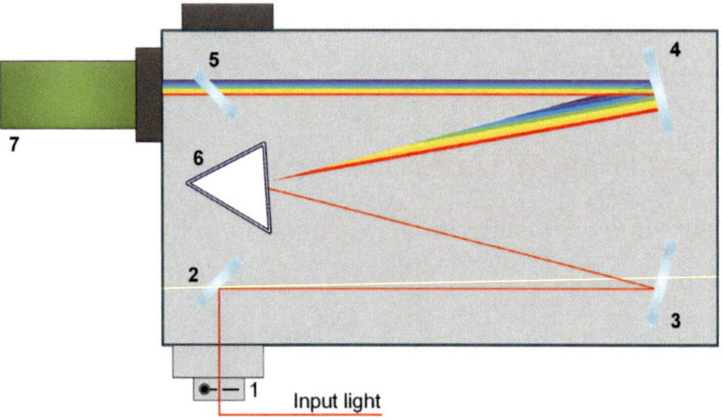


**Fig. 3.7** The schematic of a spectrograph with different parts, entrance slit (1), deflecting (2, 5) and focusing (3, 4) mirrors, grating (6), and camera (7).

An image intensifier is placed in the ICCD in front of the CCD chip to improve the sensitivity of the light detection. These photon multipliers typically contain three main components, i.e. a photocathode, a microchannel plate (MCP) and a phosphor screen. The roles of these devices are to convert the incident photons into electrons which are then multiplied electrically to a number that is dependent on the gain voltage applied to the MCP. Finally, they are converted by the phosphor screen back into the photons ready for the CCD to detect. The phosphor screen is usually coupled with the CCD by a fiber optic bundle. An advantage of ICCD is profiting from the electric shutter function, since voltage photocathode can be switched within a few ns which enables fast controllable photography.

The EMCCD cameras are alternative not using a standard image intensifier. Instead, an electron multiplying arrangement is inserted between the end of the shift register and the (output) amplifier. The design of this gain register is such that high voltage causes impact ionization which generates new electrons. Consequently, this technique enables charge from each pixel to be multiplied on the sensor before it is read out.

In a CMOS camera, image sensors convert light into electrons. In most CMOS devices, there are several transistors at each pixel that amplify and move the charge using more conventional wires. The CMOS approach is flexible because each pixel can be read individually. It converts pixel measurements simultaneously by using circuitry on the sensor itself and use separate amplifiers for fast readout. High pixel number enables a spatial resolution.

Due to the different processes applied in CCD, ICCD CMOS, and EMCCD systems, their behaviors are different regarding resolution, speed, sensitivity, and noise which should be considered for spectroscopy applications. This work profits from ICCD cameras to perform optical emission spectroscopy and absorption spectroscopy with high sensitivity. Besides, a CMOS camera is used for video cinematography and another one for video spectroscopy.

### 3.2.2 High-speed cinematography

High-speed cinematography is carried out using two types of high-speed camera (IDT-MotionPro Y6 and Y4-S3) based on CMOS technology. One of the cameras is positioned to record the arc evolution on the anode, and the other camera is installed inclined to the arc axis to record the overall discharge behavior.

The model Y4-S3 is monochromatic with a recording speed of 7000 fps at maximum resolution of $1024\times1024$ (full chip). Minimum exposure time of the device is 1 $\mu$s with a pixel depth of 10 bit. The color mode Y6 needs 3 detectors per pixel equipped with color filter and thus, lower sensitivity. It has the maximum of 1150 fps at maximum resolution of $1504\times1128$ with minimum exposure time of about 1 $\mu$s. The temporally limiting parameter is the time needed for ADC conversion and readout. Considering the diameter of electrode and gap length, the acquisition area can be optimized to achieve the higher frame rates up to 30 000 fps by reduction of the number of lines. To avoid an image-overexposing, natural density (ND) filters are integrated into the path of light or in front of the camera lens depending on the arc current. The range of transmission varies from 10% to 80% depending on the current and exposure time of the cameras.

### 3.2.3 Video spectroscopy

The video spectroscopy enables studying the temporal evolution of lines of different atoms and ions including the spatial profiles during one single pulse. Therefore, it is appropriate also for cases with limited reproducibility. The spectroscopic setup consists of a spectrograph with 0.5 m focal length (Princeton Instruments Acton SP2500) and the high speed camera IDT-MotionPro Y4-S3.

The inter-electrode gap and parts of the electrodes are imaged to the entrance slit of the spectrograph by means of either a long distance microscope or by optics based on deflecting and focusing mirrors. The obtained 2D-images contain spectral and spatial information along the arc axis as well. Fig. 3.8 represents two examples of 2D spectra captured by video spectroscopy along the arc axis together with a 1D spectra at 0.5 mm distance to the anode surface, measured during diffuse mode (top) and footpoint mode (bottom).

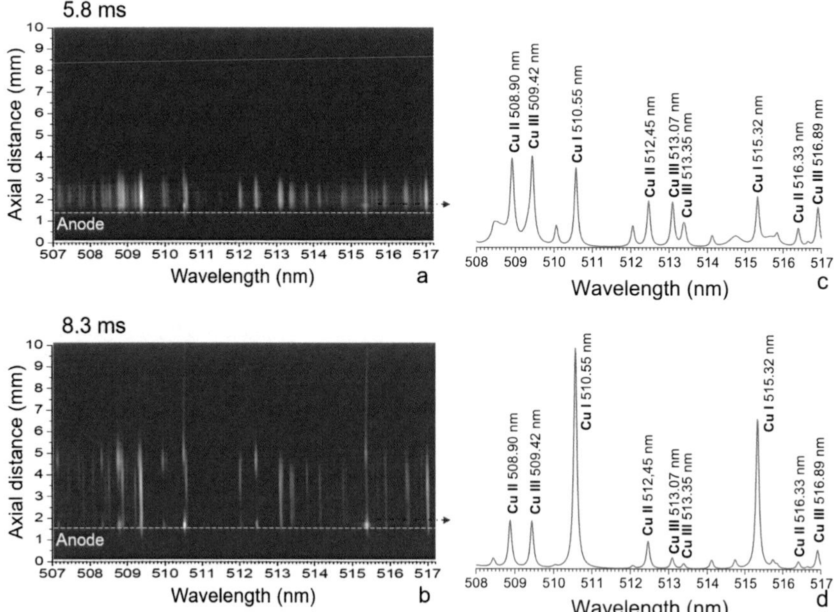


**Fig. 3.8** 2D spectra of diffuse (a) and footpoint mode (b) and the dedicated 1D spectra observed near the anode in the wavelength range 508 nm to 517 nm (c and d) [73].

In this study, entrance slits with divers widths from 30 to 50 $\mu$m are used. For conducted experiments, three different gratings with 150, 1800, and 2400 lines per mm are selected. Frame rate and exposure time are adapted depending on the experiments conditions, e.g. current magnitude, time of acquisition, etc. It should be noted that in case of video spectroscopy the spectral sensitivity of CMOS camera has to be considered, cf. Fig. 3.9. In this attempt, the wavelength is adjusted in the range of 440 – 580 nm for video spectroscopy.

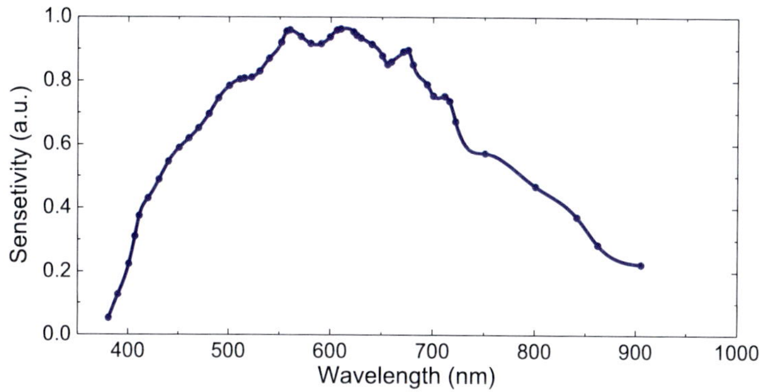


**Fig. 3.9** The sensitivity of the CMOS camera from about 400 nm to 900 nm (Motion Pro Y4).

## 3.2.4 Broadband absorption spectroscopy

The broadband absorption spectroscopy can be used to measure particle density, particularly those of ground states or of low lying excited states [74–77]. Therefore, a background light source is required with radiances at least in the order of magnitude of the test object. The background light is supplied by a pulsed high-intensity xenon lamp with a radiance of a Planck radiating of 12 000 K [78]. The pulse width of the lamp is about 1 ms with a peak power of about 1 MW.

Discharge circuit for xenon lamp is given in Fig. 3.10. Notice that the intensity should be constant during the discharge, while it should be as short as possible by rise and decay time. This condition can be fulfilled by using a pulsed DC, whose wave impedance $Z$ is matched with the resistance $R$ of the discharge in tube. The resistance of the discharge can be calculated as

$$R_e = \frac{4l_e}{\sigma \pi d_e^2} \approx 0.7\,\Omega \tag{3.16}$$

where $l_e$ and $d_e$ are the length and diameter of the lamp and equal 11 cm, 0.6 cm. $\sigma$ stands for specific conductivity which is $55\,\Omega^{-1}\text{cm}^{-1}$.

To match the wave impedance $Z$ with the resistance of the lamp $R$, the circuit elements, i.e. $L_0$ and $C_0$ and the duration of the pulse, $t$ are determined from:

$$R = (L_0/C_0)^{1/2} = Z, \qquad t = 2n(L_0C_0)^{1/2} \tag{3.17}$$

where $n$ is the number of stage and $t$ is the pulse duration. The single line diagram of discharge circuit, which is used in this work for xenon lamp, is shown in Fig. 3.11. The current waveform of discharge circuit is displayed in Fig. 3.12.

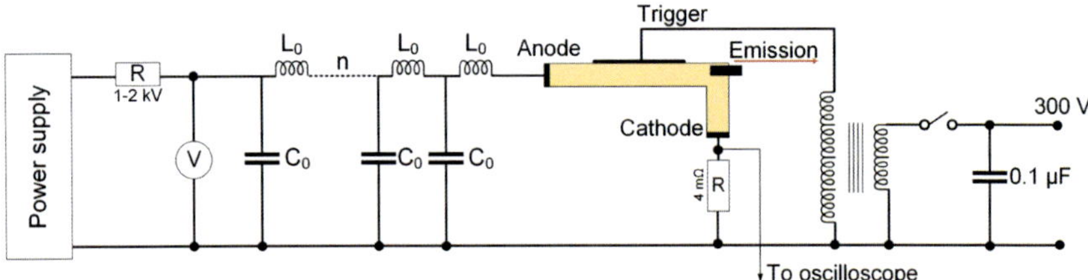


**Fig. 3.10** Discharge circuit for xenon lamp.

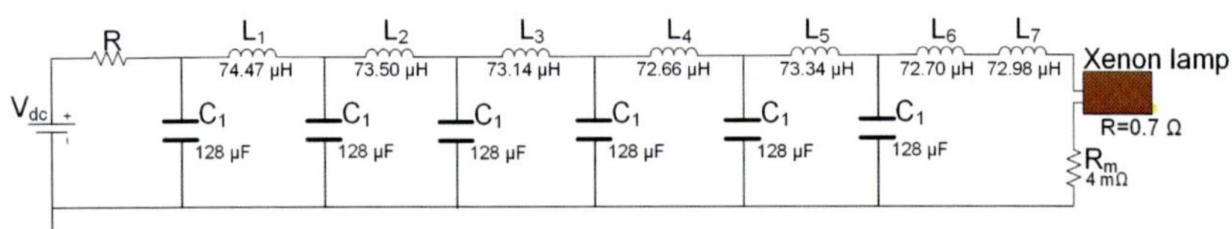


**Fig. 3.11** The single line diagram of discharge circuit which is used in this work for xenon lamp.

For absorption spectroscopy, Czerny-Turner spectrograph (Andor Technology Ltd. Shamrock 750) with 0.75 m focal length is preferred. To achieve a high spectral resolution, a grating with 1800 lines/mm and an entrance slit of 50 $\mu$m width is opted. Registration of acquired spectra is performed spatially resolved in one direction by an intensified charge coupled device (ICCD) camera (Andor Technology Ltd. iStar-DH334T-18H, 1024×1024 pixels).

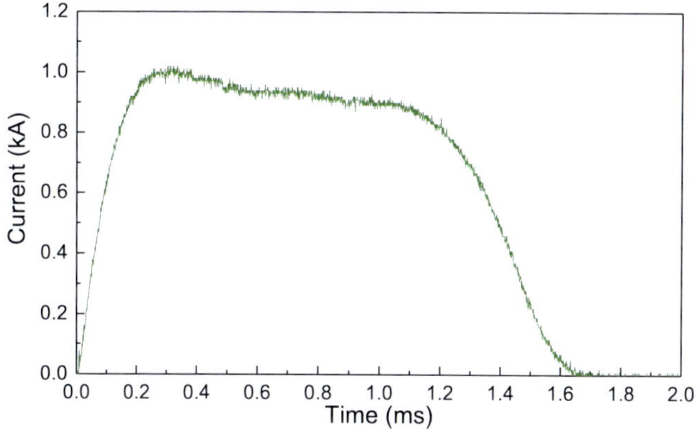


**Fig. 3.12** The current waveform of discharge circuit for xenon lamp.

## 3.3 Theories of the applied spectroscopic methods

In this section, the admitted theories to analyze the spectroscopic results and to determine the physical parameters, e.g. temperature, ground state density, emission coefficient, electron density, and etc. are introduced.

### 3.3.1 Boltzmann plot

The Boltzmann plot is a method to determine plasma temperatures from relative line intensities in optically thin thermal plasmas [79, 80]. The particle density and the pressure can also be estimated, if measured line profiles are calibrated for absolute radiance. The Boltzmann plot is based on the emission coefficient relationship. The temperature is determined from the slope of the Boltzmann plot presented in (3.18).

$$ln\left(\frac{\varepsilon^i(r)\lambda^i}{A^i_{ul}g^i_{ul}}\right) = -\frac{1}{k_B T(r)}E^i_u + ln\left(\frac{hcp/Q(T)}{4\pi k_B T(r)}\right) \tag{3.18}$$

where $\varepsilon$ is emission coefficient, $\lambda$ is the emission wavelength, $A_{ul}$ is the transition probability, $E_u$ and $g_u$ are the excitation energy and statistical weight of the excited (upper) state, respectively. $k_B$ is the Boltzmann constant, $T$ is the plasma temperature, $h$ is the Planck constant, $c$ is the speed of light, $p$ is the pressure, and $Q(T)$ is partition function. The equation (3.18) gives the approximated plasma temperature under local thermodynamic equilibrium (LTE).

In the case of two lines, this method can be simplified to an equation that directly gives $T(r)$. If the upper line energies are too close to each other, the linear fit would suffer from unacceptable uncertainties [79]. Moreover, the transition probabilities $A_{ul}$ have to be known with a sufficient accuracy. It should be remarked that the available transition probability values from different sources can significantly differ from ones claimed in some literatures. Generally, the two Cu I lines at 510.55 nm and 515.32 nm are well suited for the Boltzmann plot method due to their excitation energies including a gap energy of around 2.4 eV. Under the conditions of LTE, the plasma temperature can be directly read out from the slope of the plot. Conclusions have to be drawn carefully if LTE conditions are not more valid for plasma. In the case of partial LTE for instance, the population density of excited levels may be Boltzmann distributed, whereas the ground state is over-populated with respect to the distribution temperature of excited states [79]. Beside the temperature dependent slope of the Boltzmann plot, the axis intercept $y_0$ at $E_u = 0$ offers further information allowing to compute the pressure from the following equation:

$$p = \frac{4\pi k_B T Q(T)}{hc\, exp(-y_0)} \tag{3.19}$$

### 3.3.2 Radiating density

The reconstruction of the local radiation properties from the measured radiances typically needs a number of assumptions. Assuming rotational symmetry of the arc and radially decreasing intensity profiles, the local spectral intensity $\varepsilon(\lambda, r)$ for radial positions $r$ is related to $L(\lambda, y)$ recorded over a cross section of the arc according to the Abel transformation [81]:

$$L(\lambda, y) = 2\int_y^R \frac{\varepsilon(\lambda, r)}{\sqrt{r^2 - y^2}} dr \tag{3.20}$$

and can be obtained from Abel inversion of $L(\lambda, y)$ for each wavelength. In the following example, the spectrograph is aligned horizontally to observe the emission parallel to the anode surface. Spectra are recorded with 10 000 frames/s and exposure time of 98 $\mu$s has been singled out with spectral range 437 – 580 nm. Fig. 3.13a and c present the 2D patterns at distance 0.1 mm from the anode surface during anode spot types 1 and 2 [1].

[1] Anode spot type 1 and type 2 are introduced in 4.2.1.

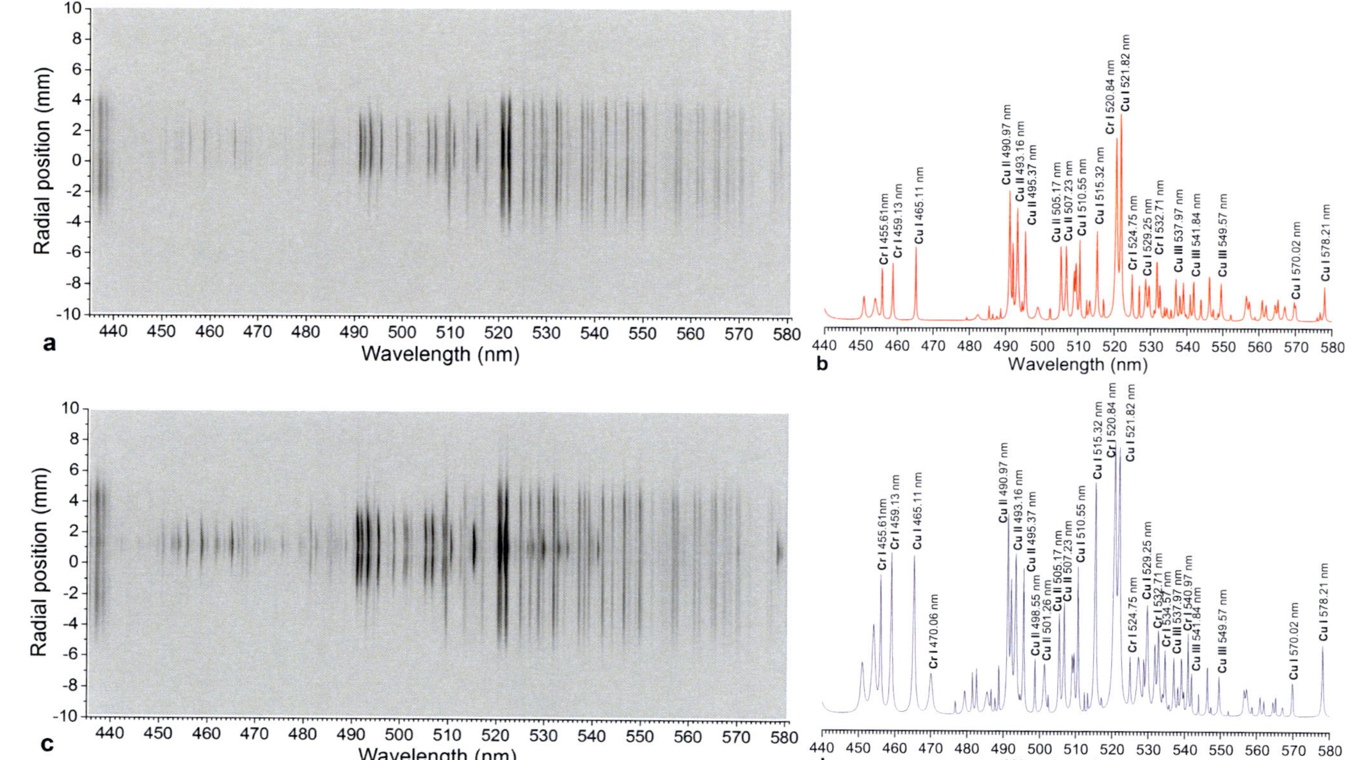


**Fig. 3.13** Spatially resolved spectra parallel to the anode surface during the formation of anode spot type 1 (top) and type 2 (bottom) with corresponding spectral profiles.

The radiating density can be determined from the local emission coefficient of a transition $\varepsilon_{lu}$ from an upper level $u$ to a lower one $l$, i.e. the density of excited Cu atoms and ions of the corresponding upper level $u$ according to [82]

$$\varepsilon_{lu} = \frac{1}{4\pi}\frac{hc}{\lambda_{lu}}n_u A_{lu} \tag{3.21}$$

where $\lambda_{lu}$ is the transition line wavelength, $A_{lu}$ is the transition probability, $h$ is the Planck constant, the speed of light is commonly denoted as $c$, and $n_u$ is radiating density.

Wavelength, lower and upper energy levels, upper level degeneracy and transition probability for the Cu I and Cu II emission lines are featured in Table 3.5 [83–86].

**Table 3.5** Wavelength, energy levels, upper level degeneracy and transition probability for the Cu I emission lines used in this work.

| **Wavelength [nm]** | **Upper energy level [eV]** | **Lower energy level [eV]** | **Upper degeneracy** | **Transition probability [$s^{-1}$]** |
|---|---|---|---|---|
| Cu I 510.55 | 3.817 | 1.380 | 4 | $2.0 \times 10^6$ |
| Cu I 515.32 | 6.191 | 3.789 | 4 | $6.0 \times 10^7$ |
| Cu I 521.82 | 6.192 | 3.817 | 6 | $7.5 \times 10^7$ |
| Cu I 578.21 | 3.786 | 1.642 | 2 | $1.7 \times 10^6$ |
| Cu II 491.78 | 17.12 | 14.60 | 9 | $2.0 \times 10^8$ |

### 3.3.3 Electron density using Stark broadening

Different methods have been applied to determine the electron density [87–90]. Stark broadening of spectral lines results from the interaction of the radiation with the electrons of the surrounding plasma leading to a broader perturbation of the lines of upper and lower levels [20]. The corresponding line width is proportional to the electron density $n_e$ and has weak temperature dependency, so that measured line widths can be used to deduce the electron density. The spectra obtained from Stark broadening has Voigt function presented in (3.22) that is a convolution of Gauss (3.23) and Lorentz (3,24) profiles.

$$V(x;\sigma,\gamma) = \int_{-\infty}^{+\infty} G(x';\sigma)L(x-x';\gamma)dx' \tag{3.22}$$

$$G(x;\sigma) = \frac{e^{-x^2/2\sigma^2}}{\sigma\sqrt{2\pi}} \tag{3.23}$$

$$L(x;\gamma) = \frac{\gamma}{\pi(x^2+\gamma^2)} \tag{3.24}$$

where $V(x;\sigma,\gamma)$ is Voigt function, $G(x;\sigma)$ and $L(x;\gamma)$ are Gauss and Lorentz functions, respectively. $x$ is the shift from the line center, $\gamma$ is the half-width at half-maximum (HWHM) of the Lorentzian profile and $\sigma$ is the standard deviation of the Gaussian profile, related to its HWHM. An example of a 2D pattern together with the corresponding spectral profiles at 0.1 mm distance from the anode surface during anode spot is presented in Fig. 3.14.

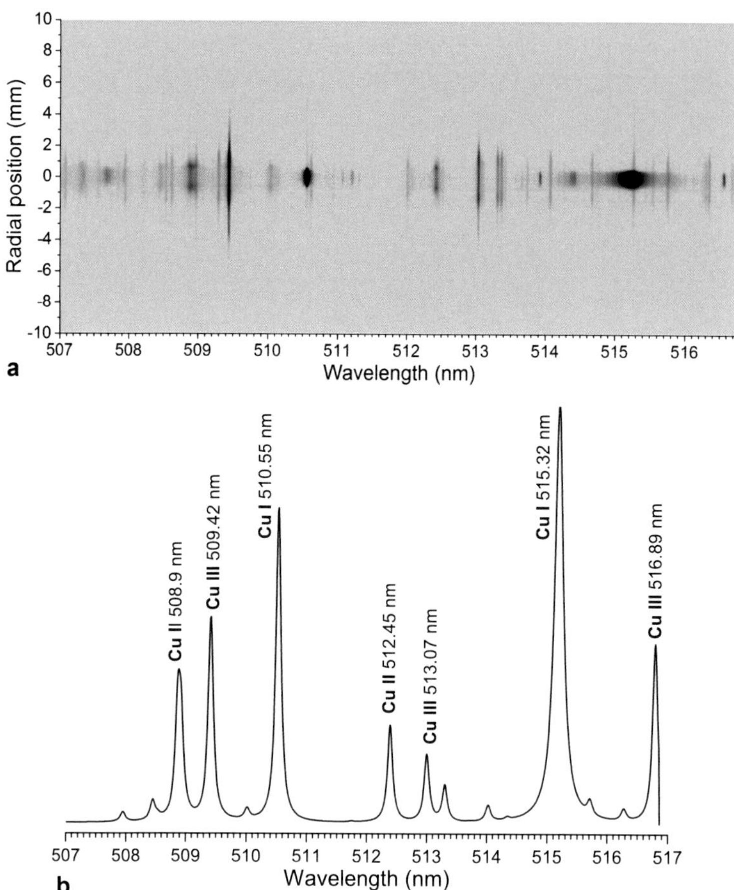


**Fig. 3.14** 2D patterns together with the corresponding spectral profiles at 0.1 mm from the anode surface during anode spot type 2.

The Gauss contribution can be identified with the instrumental profile of the spectroscopic system. For this purpose, the instrumental profile of the spectrometer is measured using a commercial low-pressure Hg-Ar discharge tube (Penray).

To measure the temporal profile of electron density, the video spectroscopy technique is applied. The grating of 1800 l/mm and width of entrance slit of 40 $\mu$m are chosen. The central wavelength of 510 nm is opted to capture both Cu lines at 510.55 nm and 515.32 nm. Thereby, temporal evolution of the electron density could be disclosed through Cu I 515.32 nm. The instrumental profile of the applied spectrograph based on the Gauss fit is plotted in Fig. 3.15. The width of Gauss profile equals 0.039 nm. The Lorenz contribution of Voigt function determines the width of Stark broadening [20]. Different methods are proposed to reveal the electron density based on the Stark broadening.

In [20] a Stark width of $\Delta\lambda_0$=0.19 nm (FWHM) can be found for Cu I at 515.32 nm an electron density of $n_{e0}=1\times10^{23}$ m$^{-3}$ and the temperature of $T_0$=10 000 K. The following formula can be employed to estimate the electron density, $n_e$.

$$\frac{\Delta\lambda}{\Delta\lambda_0} = \frac{n_e T^{1/6}}{n_{e0} T_0^{1/6}} \tag{3.25}$$

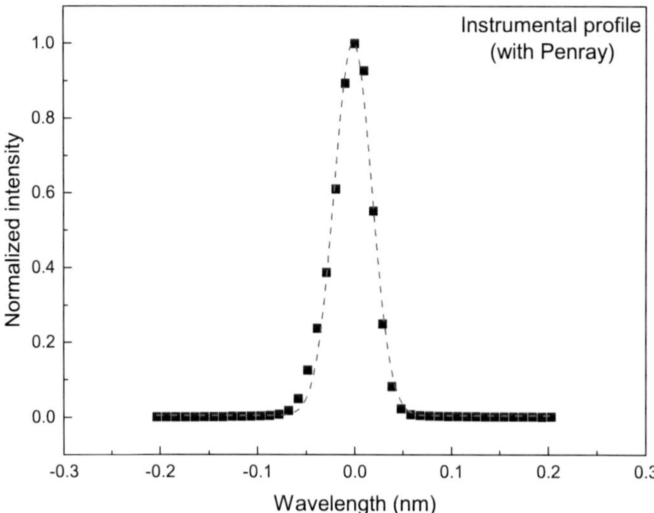


**Fig. 3.15** The instrumental profile of the applied spectrograph based on the Gauss fit.

The measured values and Voigt fit for Cu I 515.32 nm are visualized in Fig. 3.16 at two time instants during formation of anode spot. The electron density for anode spot type 1 and type 2 are $8\times10^{22}\,\mathrm{m}^{-3}$ and $4.3\times10^{22}\,\mathrm{m}^{-3}$, respectively.

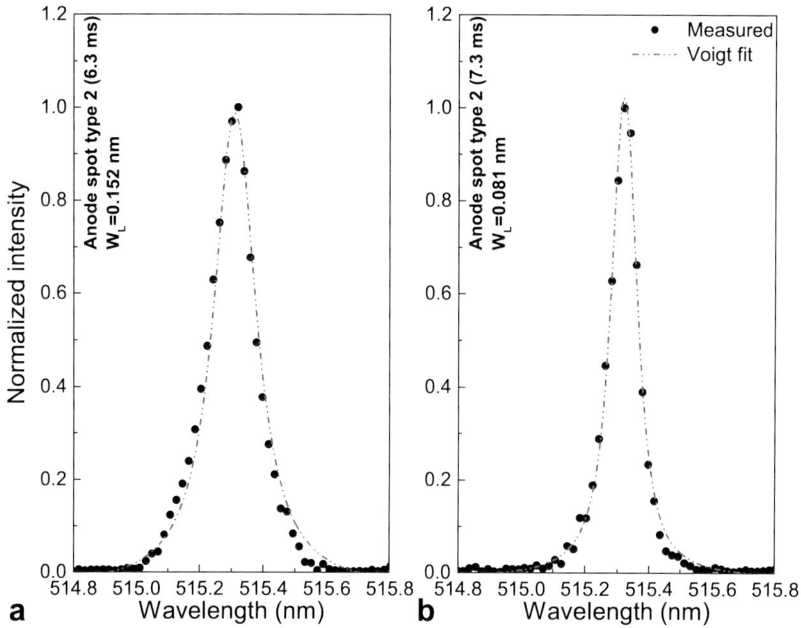


**Fig. 3.16** The measured (symbols) and Voigt fit (dashed-dotted-dotted line) for Cu I 515.32 nm at anode spot type 2.

## 3.3.4 Ground state density using absorption spectroscopy

The absorption coefficient is related to the lower state population density $N_l$ of the transition expressed in the following formula [76]

$$\int_0^\infty kDd\lambda = \int_0^\infty ln(\frac{L_\lambda}{PL_\lambda - P_\lambda})d\lambda = \frac{\pi e^2 \lambda_0^2 D N_l f_{lu}}{4\pi\varepsilon_0 m_0 c^2} \tag{3.26}$$

where $\varepsilon_0$ denotes electric constant, $m_e$ electron mass, $c$ the speed of light in vacuum, $e$ elementary charge, $f_{lu}$ the oscillator strength , $\lambda_0$ the center wavelength of the considered line. $L_\lambda$ stands for the background light source (measured without plasma), $PL_\lambda$ for the radiation of the plasma and the background light source, $P_\lambda$ for the plasma radiation (measured without background), $k(\lambda)$ for the absorption coefficient and $D$ for the effective absorption length. From optical observation and 2D spectroscopy, it is estimated to be about 10 mm in this work. The left term of (3.26) can be determined from the transmission coefficient [76]

$$T(\lambda) = \frac{L_\lambda}{PL_\lambda - P_\lambda} \tag{3.27}$$

For particle density estimation, three Cr resonance lines are evaluated (cf. Table 3.6). Using the Lambert-Beer's law, the absorption coefficient $k$ can be obtained from transmittance $T$ by [76]

$$T(\lambda) = exp[-k(\lambda)D] \tag{3.28}$$

In this effort, the absorption spectroscopy is conducted for both high-current phase and after current zero. Intensity of the plasma light could be overlooked after current zero. Here, the intensity of the background source is high enough to register a clear absorption signal [76]. Consequently, the transmission $T$ in (3.24) after current zero can be derived by [76, 77]

$$T = \frac{I}{I_0} \tag{3.29}$$

where $I$ symbolizes the transmitted spectral intensity and $I_0$ is the spectral intensity outside the absorption profile (Fig. 3.17). It is assumed that there is no absorption by the plasma among the observed spectral lines.

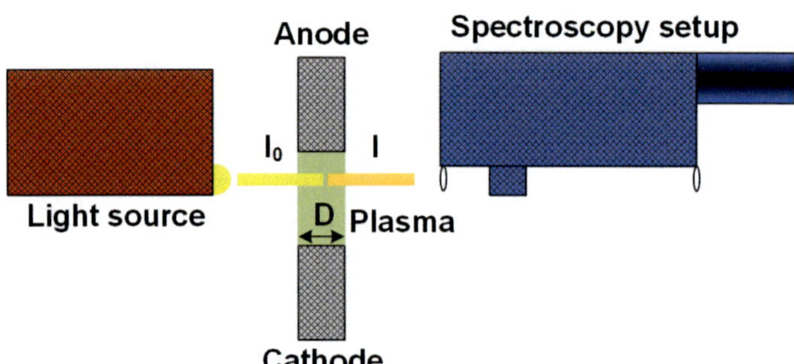


**Fig. 3.17** Optical arrangement setup corresponding to the spectrograph for absorption spectroscopy after current zero.

A spectral range between 424 and 431 nm is chosen to study the Cr I density. It contains spectral lines of three resonance transitions of Cr I at 425.43 nm, 427.78 nm, and 428.97 nm. Spectroscopic constants for the selected lines including wavelength, transition probability ($A_{ki}$), and oscillator strength ($f_{lu}$) can be found in Table 3.6.

An example of a 2D spectrum captured by Czerny-Turner spectrograph at about 100 $\mu$s after current zero is visualized in Fig. 3.18a.

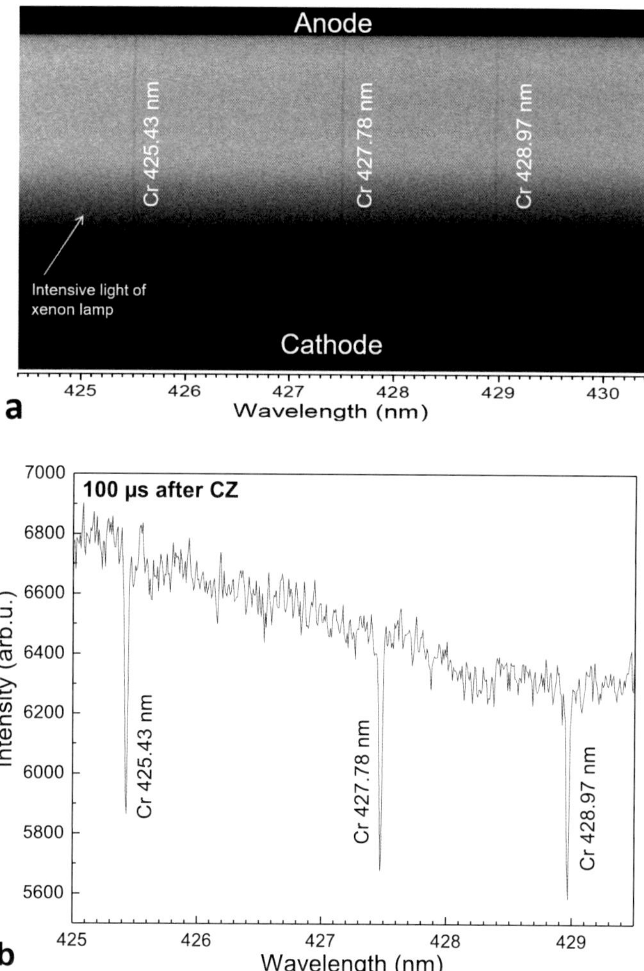


**Fig. 3.18** a) 2D patterns together with the corresponding spectral profiles at 0.1 mm from the anode surface during anode spot type 2 and b) The relative intensities of the plasma transmission and the background light source integrated along the wavelength [76].

The gray background between anode and cathode corresponds to the light emitted by xenon lamp, in which three thin vertical lines could be distinguished at 425.43 nm, 427.78 nm, and 428.97 nm indicating three resonance lines of Cr I.

Fig. 3.18b presents the relative intensities of the three absorbed resonance lines integrated along the wavelength corresponding to Fig. 3.18a. The signal is averaged over a horizontal stripe of approximately 4 mm in axial direction in front of anode.

Transmission can be found using (3.29) proposed for Cr I at 425.43 nm in Fig. 3.19.

**Table 3.6** Atomic constants for the applied Cr I resonance lines.

| **Wavelength [nm]** | **Transition** | $\mathbf{f_{lu}}$ | $A_{lu}$ **[1/s]** |
|---|---|---|---|
| 425.43 | $3d^5(6^s)4s \rightarrow 3d^5(6^s)4p$ | 0.11 | $3.15 \times 10^7$ |
| 427.48 | $3d^5(6^s)4s \rightarrow 3d^5(6^s)4p$ | $8.42 \times 10^{-2}$ | $3.07 \times 10^7$ |
| 428.97 | $3d^5(6^s)4s \rightarrow 3d^5(6^s)4p$ | $6.23 \times 10^{-2}$ | $3.16 \times 10^7$ |

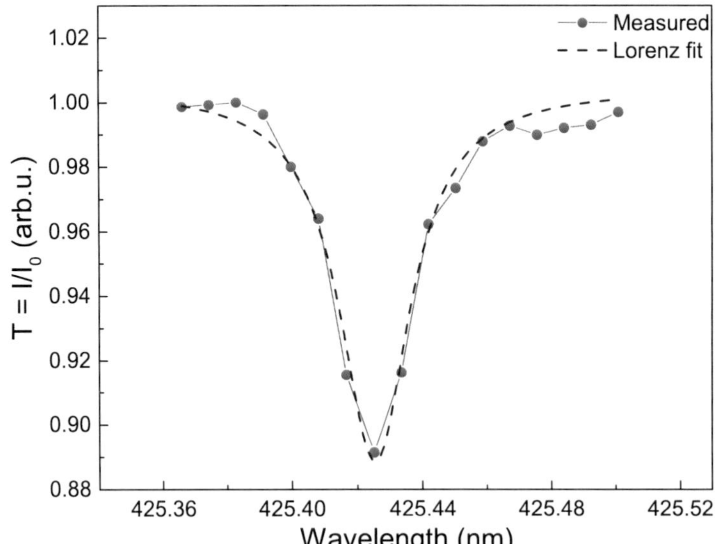


**Fig. 3.19** An example of transmission curve $T$ for Cr I 425.43 nm (circles) along with the line fit (dashed line) [76].

The focus of the present study is to designate the chromium lines providing better optical accessibility (visible light range) than the copper lines (UV range). The general behavior of the copper component is expected to be very similar to that of chromium being revealed in [15] for the active phase of an arc discharge with similar electrodes.

In the case of high-current phase, (3.27) must be applied; the radiation of the plasma and the background light source, the plasma radiation as well as the background light source must be captured. One spectrograph is applied to absorb the radiation of plasma and background light at the same time. Therefore, two optical paths are aligned to the spectrograph slit as demonstrated in Fig. 3.20.

The first one captures the plasma radiation only (mirror 2) and the second one the radiation of both plasma and lamp (mirror 1). It is noteworthy to mention that the spectrum of the background light source is measured separately without discharge. Accordingly, its stability is investigated in terms of shot to shot variation and the relation of signals.

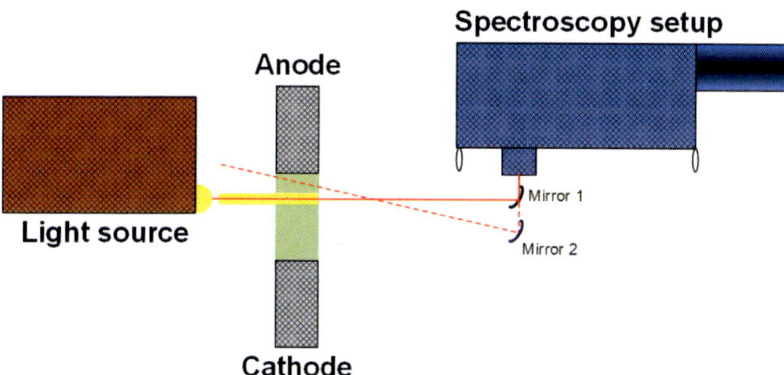


**Fig. 3.20** Optical arrangement setup corresponding to the spectrograph for absorption spectroscopy during high-current modes.

An example of a 2D spectra of the background light source and plasma that of the plasma only presented in Fig. 3.21a and b, respectively. The gray background corresponds to the light emitted by xenon lamp. As it can be seen, the radiation of background light source is more intense than the radiation of plasma. The three thin vertical lines at 425.43 nm, 427.78 nm, and 428.97 nm are associated to the three resonance lines of Cr I.

The relative intensities of the plasma transmission and the background light source corresponding to Fig. 3.21a and b are integrated along the wavelength and presented in Fig. 3.21c. The signal is averaged over a horizontal stripe of approximately 5 mm in axial direction in front of the anode. Transmission and density of Cr can be calculated using (3.27) and (3.26), respectively.

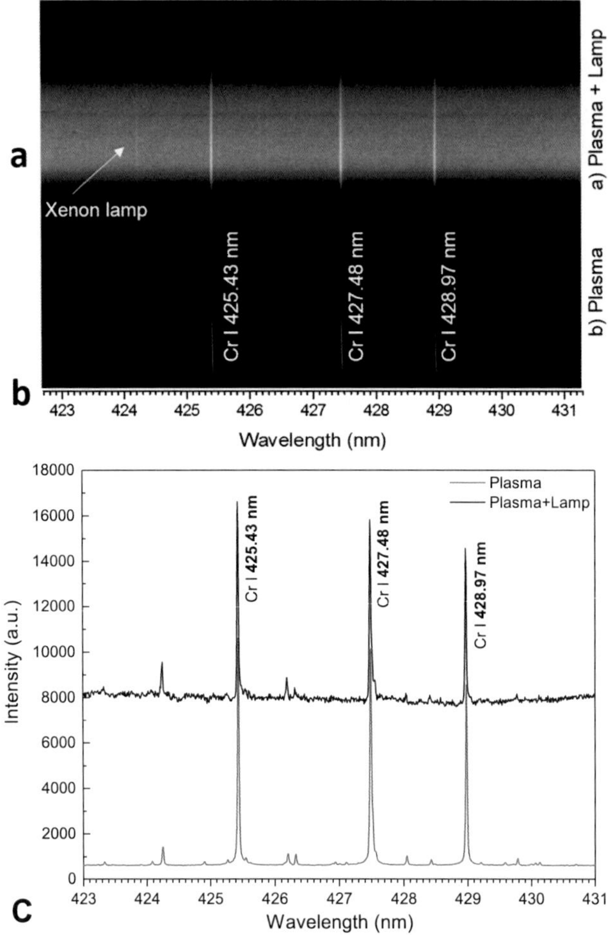


**Fig. 3.21** a) An example of a 2D spectrum with three Cr I lines during high-current anode modes and b) The relative intensities of the plasma transmission and the background light source integrated along the wavelength during high-current anode modes.

## 3.4 Calibration

To quantitatively characterize those physical specifications of plasma in 3.2, absolute calibration side-on spectra in units of spectral radiance is necessary.

This is realized by placing a tungsten strip lamp (OSRAM Wi 17/G) with known spectral radiance at defined current. In this regard, the same slit width is opted and the distance is adjusted to the same as that of the arc position. Hereby the spectral radiance can be calculated as follow

$$L_{\lambda P} = \frac{B_{\lambda}}{S_{\lambda WBL}} S_{\lambda P} \left( \frac{t_{WBL}}{t_P} \right) \tag{3.30}$$

where $B_{\lambda}$ stands for the calibration curve of tungsten strip lamp (in unit of spectral radiance, $\mathrm{Wnm^{-1}m^{-2}sr^{-1}}$), $S_{\lambda WBL}$ is the measured signal of tungsten strip lamp and $S_{\lambda P}$ is the measured signal of plasma (in counts). $t_{WBL}$ and $t_P$ denote the exposure times of the measured signals of tungsten strip lamp and plasma, respectively.

# Chapter 4

# Impact of vacuum interrupter properties

## 4.1 Introduction

The transition either from low-current mode to high-current mode or between high-current modes is governed mainly by current waveform, gap geometry, contact material, and contact opening speed [9, 12, 91, 92]. In this chapter, the different discharge modes are investigated with respect to high-speed camera images and current and voltage waveforms. To reduce the complexity of the system that is caused by the magnetic field, a pair of butt electrode with diameter smaller than 25 mm is employed without external magnetic field. However, the applied high current densities are comparable to those which exist in high-current vacuum interrupters equipped with TMF and AMF systems [1, 2]. In the following section, firstly the general behavior of such high–current modes is described exemplary for a standard case with AC pulses of 50 Hz and CuCr7525 electrodes. Then, the variation of different parameters is examined as listed in the following and in Table 4.1.

- AC pulses of 50, 180, and 260 Hz[1] and pulsed DC over 5 ms and 10 ms.
- Electrodes made out of CuCr7525 (10 mm diameter), CuCr50 (20 mm) and Cu (25 mm).
- Contact opening speed of 1 m/s and 2 m/s.

[1] It should be noted that AC pulses of 100 Hz are described in chapters 6 and 7.

- Opening time from 0 to 2 ms.

Please note that other electrodes are only used at DC pulse length variation (section 4.3.3) and for the direct comparison of different contact materials (section 4.4).

**Table 4.1** Summary of parameters variation applied in this chapter.

| **Electrodes** | **Diameter [mm]** | **Waveform** | **Opening speed [m/s]** | **Figure number** |
|---|---|---|---|---|
| CuCr7525 | 10 | AC 50 Hz | 1 and 2 | 4.1 - 4.8<br>4.16 |
| | | AC 180 Hz<br>AC 260 Hz | | 4.6 - 4.8 |
| | | DC 10 ms | | 4.9 - 4.11<br>4.15 |
| | | DC 5 ms | | 4.17 |
| CuCr50 | 20 | DC 5 ms | 1 and 2 | 4.17 |
| Cu | 25 | DC 10 ms<br>DC 5 ms | 1 and 2 | 4.12 - 4.14<br>4.14, 4.17 |

# 4.2 AC pulses

## 4.2.1 Current, voltage, and high-speed camera images

In case of AC pulses, the experiments are carried out using CuCr7525 electrodes with a diameter of 10 mm. The applied AC 50 Hz current to the test object together with the resulting voltage for opening time of 1.2 ms is plotted in Fig. 4.1. The maximum current is about 4.3 kA and the arc voltage varies between about 20 and less than 40 V. The plot discloses also the gap length, i.e. the distance between the electrodes, as a function of the time (traveling curve). In the example, the gap opening starts at about 1.2 ms, i.e. at nearly 2 kA. The traveling curve is nearly linear; an averaged contact opening speed of about 1 m/s can be deduced. From Fig. 4.1 and the according images captured by video camera, several discharge modes can be distinguished. This includes diffuse mode, footpoint [1], and anode spot type 1 which are time resolved to estimate the approximate transition time of the discharge. Notably, the characteristics of the anode spot type 1 is the same as the anode spot which is presented by Miller in [7–11]. However, two new high-current anode modes have been found recently, i.e. spot type 2 and anode plume. They can be formed after the anode spot type 1 under certain conditions, e.g. higher current and gap length [73]. The characteristics of anode spot type 2 and anode plume will be presented later but it should be pointed out that due to the differences of known phenomena a new classification is required. Therefore, the anode spot defined by Miller is named here as anode spot type 1 [73, 92].

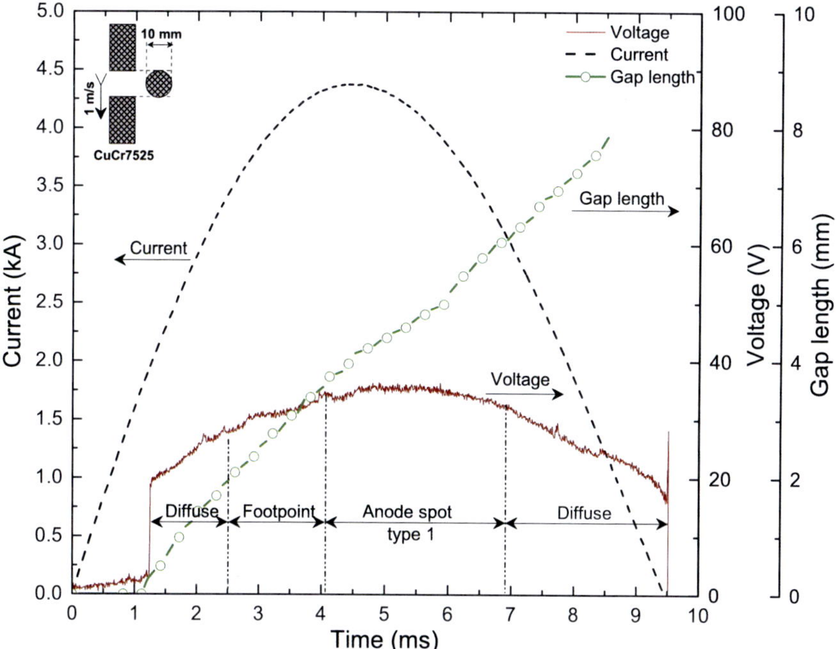


**Fig. 4.1** The current and voltage waveforms and the electrode traveling curve for AC 50 Hz. The different high-current anode modes including footpoint, and anode spot type 1 are indicated between dashed-dotted lines. Contact material: CrCr7525, diameter of electrodes: 10 mm, contact speed: about 1 m/s, opening time: 1.2 ms.

Fig. 4.2 presents selected discharge states captured by the high-speed camera at different episodes corresponding to the identified discharge modes in Fig. 4.1.

[1] Although the correct notations are diffuse mode, footpoint mode, and anode spot mode, for simplification diffuse, footpoint, and anode spot are used, respectively.

In each frame, the anode and cathode are at the top and bottom, respectively. A diffuse mode can be observed at the beginning of the discharge (Fig. 4.2a). The first high-current anode mode, footpoint, is found at the time instant of 2.5 ms, i.e. little more than 1 ms after arc ignition and at a current of about 3.5 kA. After about 4 ms and near to maximum current, the anode spot type 1 appears which can be detected from the high bright area on the anode (Fig. 4.2d). During the footpoint and the anode spot type 1, the average arc voltage increases by 15 V. The anode spot type 1 continues up to about 6.8 ms and then it transforms back to the diffuse mode at the time 7 ms as depicted in Fig. 4.2h. It should be noted that the current is below 3 kA and the gap length more than 6 mm at this time.

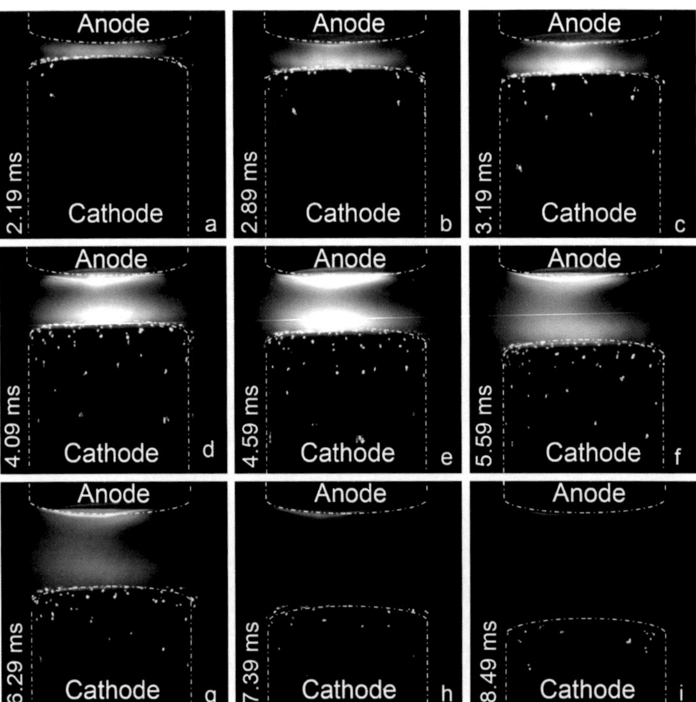


**Fig. 4.2** The appearance of diffuse mode (a and i), footpoint (b and c), anode spot type 1(d to g). The corresponding onset times for different discharge modes are presented.

Under nominally similar conditions (except arc opening time), also the formation of an anode spot type 2 following the anode spot type 1 may occur. The current and voltage waveforms together with the traveling curve for opening time of 0.3 ms are presented in Fig. 4.3. High-speed camera images corresponding to Fig. 4.3 are compiled in Fig. 4.4.

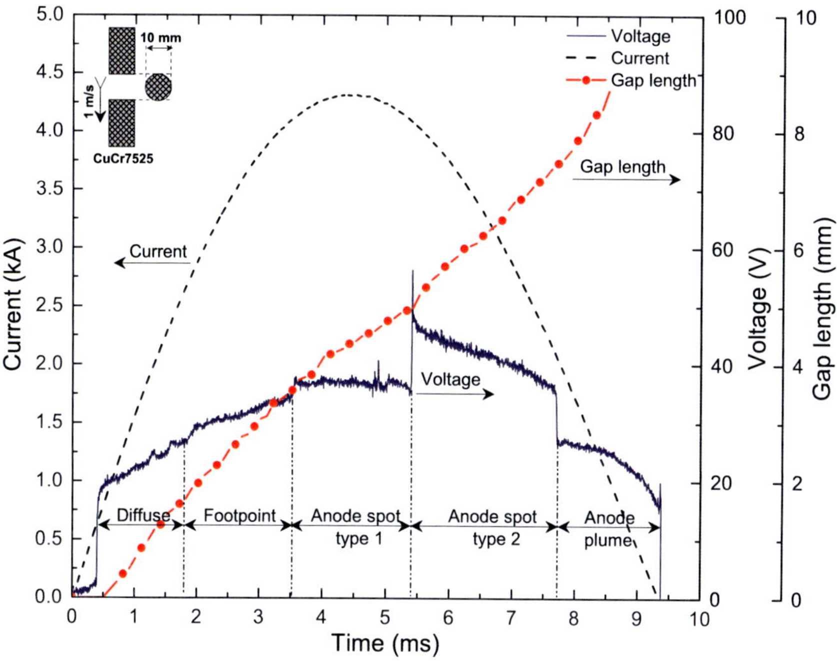


**Fig. 4.3** The current and voltage waveforms and the electrode traveling curve for AC 50 Hz. The different high-current anode modes including footpoint, anode spot type 1 and 2, and anode plume are indicated between dashed-dotted lines. Contact material: CrCr7525, diameter of electrodes: 10 mm, contact speed: about 1 m/s, opening time: 0.3 ms.

The two types of anode spot modes differ in the voltage course and the radiation intensity near the anode and the cathode [73]. The differences are scrutinized in chapter 9 to comprehensively understand the reason of formation of anode spot type 2. The anode spot type 2 shows similarities with the intense mode that is characterized by strong interaction of both electrodes, but the anode spot type 2 occurs at much larger gap lengths. In the intense mode, large spots occur at both electrodes which are brighter than those in the anode spot type 1. Nevertheless, the intense arc mode normally emerges at lower electrode distances compared to the anode spot type 1; the arc voltage remains low and without noise but it is still higher than in the diffuse mode [9]. These conditions are not fulfilled here.

In case of the anode spot type 1, a considerable anode activity with one large or several smaller and very bright spots is observed [9]. Typically, it is accompanied by a change in the average arc voltage and small fluctuations in the voltage signal [9]. However, an abrupt change in the arc voltage is recorded during the transition to anode spot type 2. In Fig. 4.3, the arc voltage jumps abruptly from about 37 V at least by 10 V. This is observed at larger gap length while both anode and cathode are active.

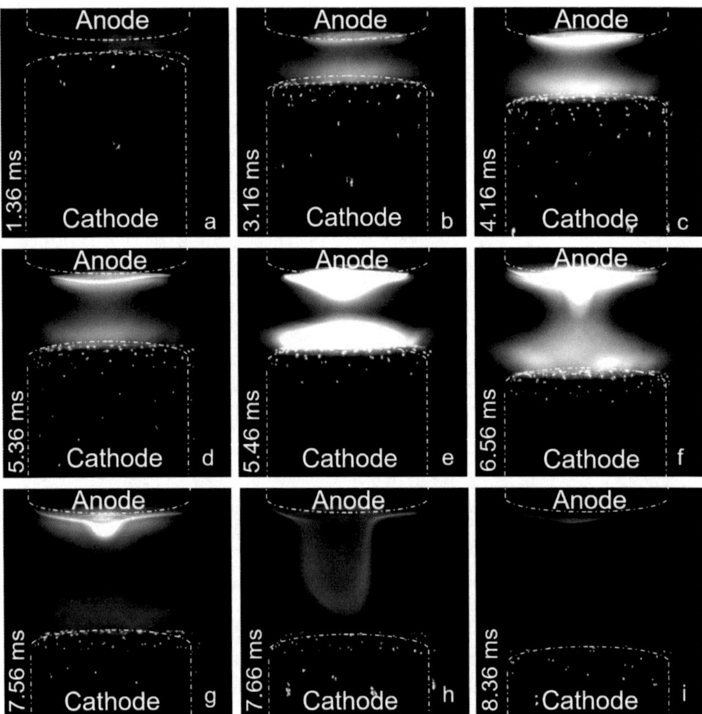


**Fig. 4.4** The appearance of diffuse mode (a, i), footpoint (b), anode spot type 1(c, d), anode spot type 2 (e to g), and anode plume (h). The onset time for each discharge mode is presented.

The approximate time of appearance of the different discharge modes is determined based on the arc voltage characteristics and high speed camera images (cf. Fig. 4.3 and Fig. 4.4). The footpoint, anode spot type 1, and anode spot type 2 are started at 1.8 ms, 3.5 ms, 5.4 ms, respectively. The anode plume is formed after the extinction of anode spot type 2 by an abrupt decrease in the arc voltage (cf.

Fig. 4.4h). The gap length regarding each high-current anode mode is indicated in Fig. 4.3.

During the formation of anode spot type 1, the bright area near the anode progresses initially and then is dimmed in the course of time (see Fig. 4.4b and d) [73, 92]. At about 5 ms, the anode spot type 2 with high emission commences on both anode and cathode as seen in Fig. 4.4e. In the anode spot type 2, both anode and cathode glitter brighter than in the case of anode spot type 1. The abrupt increase in arc voltage (around 5 ms, see Fig. 4.3) can be attributed to the fact that the anode jet and the cathode jet collide in the gap [73]. Rather unusual for a vacuum arc, an arc column is built up with considerable field strength. During the anode spot type 2, the arc voltage continuously decreases under decreasing current, nearly reaching the level at the end of anode spot type 1 of about 37 V. After the extinction of anode spot type 2 and before zero current, the voltage abruptly drops by another 10 V. Now, an anode plume is formed (cf. Fig. 4.4h) [73]. Detailed analysis of anode plume is provided in chapter 6.

### 4.2.2 Existence diagram of different discharge modes

Usually, the parameters namely voltage, current, and gap length are measured and plotted versus time. A common alternative representation is visualization of the gap length as a function of the current, through which appeared discharge modes could be compared [9, 12, 92]. A displacement curve, representing the gap distance as a function of the current for Fig. 4.3, is drawn in Fig. 4.5a. The different discharge modes can be discriminated at specific gap lengths and corresponding currents. The onsets are marked in the graph. As it can be seen, the first high-current mode is footpoint at the current of about 2.6 kA. Thereupon, anode spot type 1 emerges followed by anode spot type 2 at current of about 4.2 kA but at different gap lengths. The last mode, anode plume, emerges at the current of 2 kA and gap length of about 7.7 mm.

To build a good understanding of formation of those abovementioned modes as a function of current and gap length, the experiments are extended and carried out at various contact opening times, opening speeds, and maximum currents. A series of 6 curves are shown in in Fig. 4.5b. The displacement curve of the abovementioned example is number 2. Curve 5 has a later gap opening (at 2.5 kA) whereas 1 and 3 are cases of earlier gap opening, lower peak current of 3.5 kA, and higher opening speed etc.

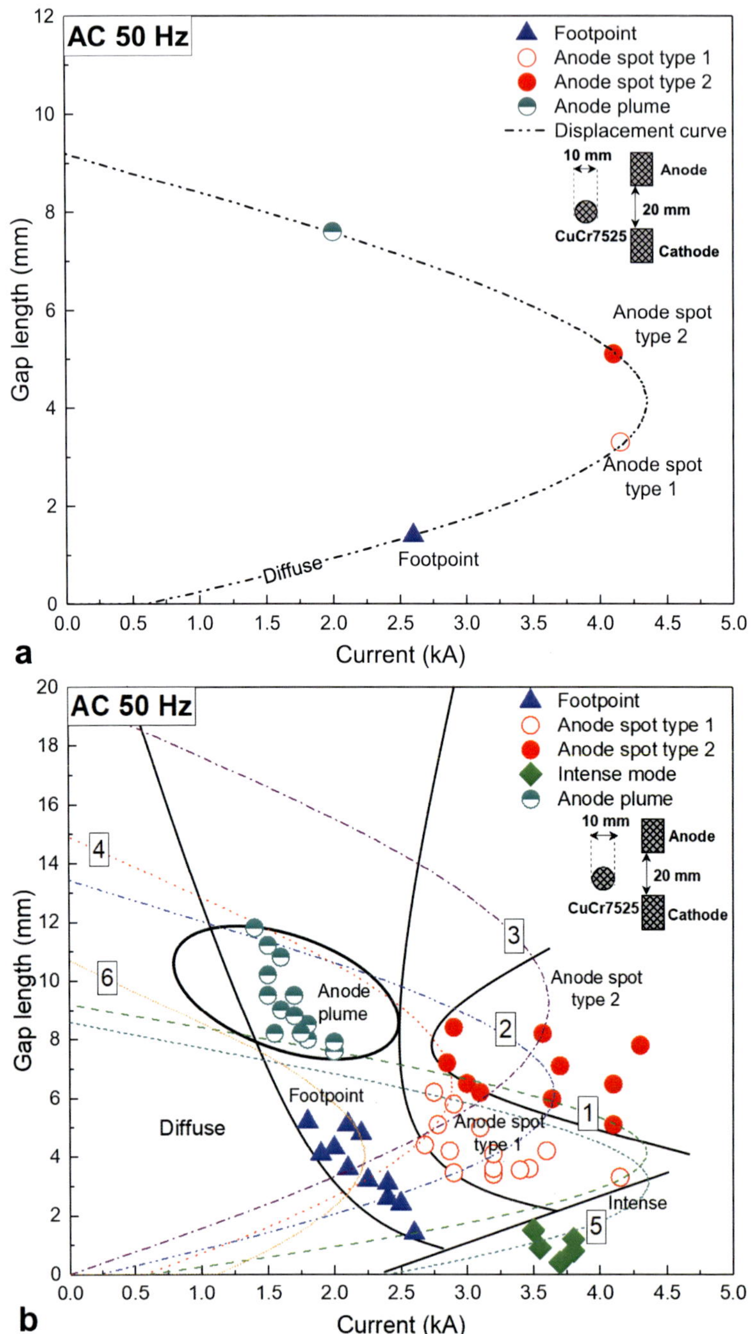


**Fig. 4.5** A displacement curve (a) and the existence diagram (b) for applied AC 50 Hz for CuCr7525 electrodes with diameter of 10 mm.

The detected modes, i.e. footpoint, anode spot type 1, anode spot type 2, intense mode, and anode plume are symbolized by filled triangles, unfilled circles, filled circles, filled diamonds, and half-filled circles, respectively as seen in Fig. 4.5a and b. Thus, an existence diagram can be drawn.

A footpoint may occur about before or after current maximum. If the footpoint appears after current maximum, then the threshold current would be lower than the corresponding maximum current and the gap length would be higher. These two types of footpoint occurrence can be used to determine the borders of footpoint-diffuse and footpoint-anode spot type 1 [92]. Anode spot type 2 appears at higher gap length but it can be observed even at lower current compared to anode spot type 1 [73, 92].

The borders between intense mode, anode spot type 2 and the other modes are assigned approximately regarding the threshold currents and corresponding gap lengths of the intense mode and anode spot type 2. It could be noticed that by increasing the number of tests, the boundaries can be simply identified. The detailed information regarding distinction of boundaries is provided in [92].

There is a considerable similarity between the regions of the existence diagram in this work compared to the works presented in [8, 9] for AC 50 Hz. It should be also noted that some parameters such as contact material, electrode diameter, and frequency can change the shape and the borders of the existence diagram [9–11].

### 4.2.3 Frequency variation of AC pulses

The high-current anode phenomenon in case of AC pulses is investigated for three different frequencies of 50, 180, and 260 Hz. The currents, voltages, travel curves, and corresponding high-speed camera images are depicted in Fig. 4.6 and Fig. 4.7, respectively. The contact speed is chosen to be either 1 or 2 m/s for those frequencies.

The peak current is about 2.85 kA for both 50 Hz and 180 Hz. The first high-current mode is the anode spot type 1 which begins around maximum current of 2.85 kA and 2.7 kA for 180 Hz and 50 Hz, respectively. Notably, the transition to or between high-current modes normally leads to increase in fluctuation of the arc voltage in cases where electrode are fabricated from Cu [92]. Nevertheless, the arc voltage can be also without any fluctuation in case of CuCr electrodes [9]. In such cases the evaluation of the light recorded by the high speed camera is the key to determine the transition to high-current modes. Therefore, to identify the time instants of high-current anode modes presented in Fig. 4.6, the light recorded by high speed camera is assessed and the time of the oscilloscope traces is tracked.

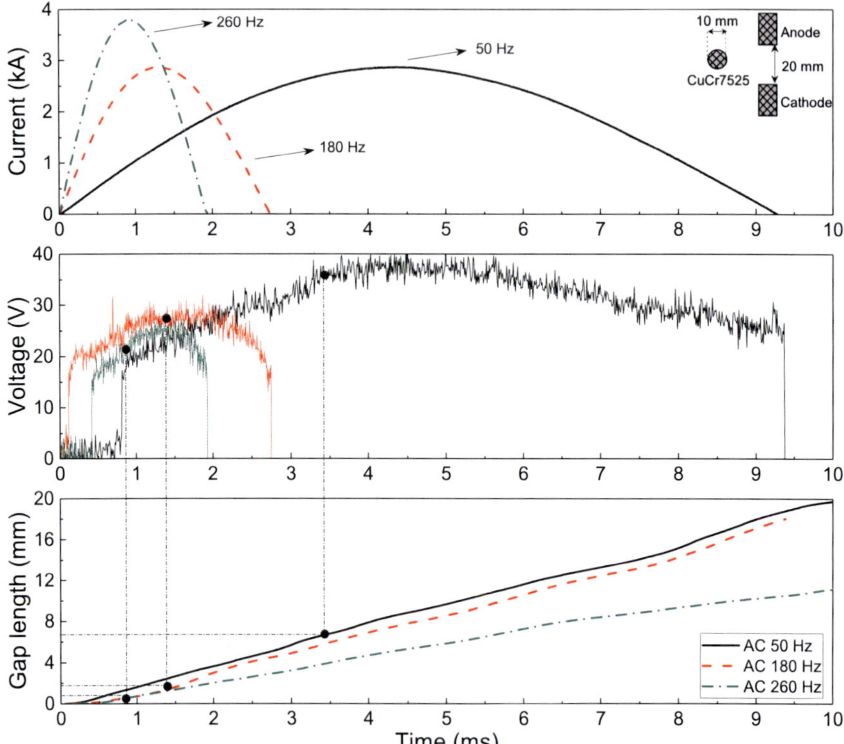


**Fig. 4.6** Current, voltage and corresponding traveling curves for different AC frequencies. Contact opening speed of 1 m/s and 2 m/s is applied [92].

By increasing the frequency to 260 Hz, the threshold current of the first high-current mode which is an intense mode is increased to 3.7 kA.

Regarding the gap length, the first high-current anode mode appears at a gap length of 6.7 mm, 1.8 mm, and 1 mm for 50 Hz, 180 Hz, and 260 Hz, respectively, being confirmed by the light emission near the anode in the high speed images (cf. Fig. 4.7). It should be noticed that by increasing the opening speed in case of high frequency currents the first high-current mode can be shifted to the anode spot type 1 or footpoint mode, which is in good agreement with [9] where the intense arc mode occurs at shorter gap lengths than the anode spot type 1.

Fig. 4.7 represents high speed images of the formation of the anode spot type 1 and intense mode for the three frequencies. In the intense mode very bright luminosity is observed that covers both anode and cathode and the inter-electrode gap as well [92]. The arc voltage during intense arc mode is always low and with small fluctuations but higher than that of in diffuse mode [92]. It should be noted that after occurrence of the high-current anode modes, the average of arc voltage rises for all investigated frequencies (see Fig. 4.6).

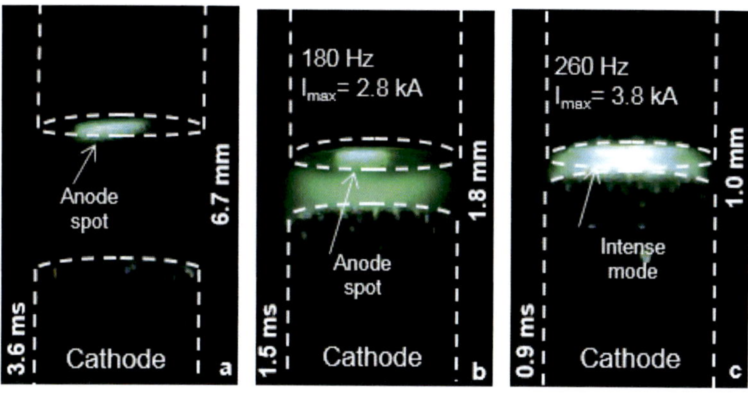


**Fig. 4.7** High speed camera images of high-current anode mode for different AC frequencies. Anode spot type 1 in case of 50 Hz (a), anode spot type 1 in case of 180 Hz (b), and intense mode in case of 260 Hz (c) [92].

A larger number of experiments are conducted at three different frequencies with varying opening speed 1 and 2 m/s, opening time (cf. 4.2.4), and peak current from 2.0 kA to 4.5 kA. The results of the threshold currents are summarized in Fig. 4.8.

In case of 50 Hz, both footpoint and anode spot type 1 are formed at both contact speeds, whereas the anode spot occurs at higher currents and gap lengths. Besides, the threshold current is higher as the contact speed is lower.

By increasing the frequency to 180 Hz and 260 Hz, the threshold current at both contact speeds is raised and the first high-current anode mode at the contact speed of 1 m/s is intense mode. Nevertheless, the anode spot type 1 with lower threshold current appears by increasing the contact speed.

Notice, the distance in which the high-current mode appears decreases extremely with ascending frequency, i.e. with a faster approach to a threshold current. This confirms that the formation of high-current anode modes is mainly influenced by the instantaneous current in case of AC pulses especially for current waveforms with higher frequencies [9, 91, 92].

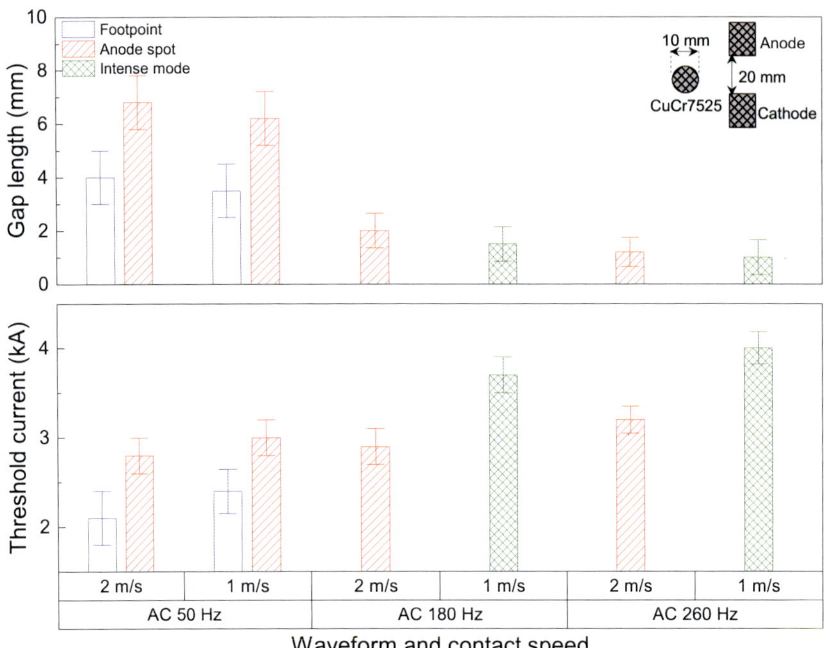


**Fig. 4.8** Threshold current of high-current anode phenomena for different AC frequencies and varying peak current from 2.0 kA to 4.5 kA. Contact opening speed of 1 m/s and 2 m/s is applied [92].

### 4.2.4 Contact opening time

The impact of arcing time on the formation of high-current anode modes is investigated by application of AC 50 Hz. Notably the arcing time proceeds at least for the whole period between one half-cycle and zero for a real circuit breaker [92]. The arcing time can be expected to have the biggest impact on the formation of high-current anode modes [9, 92]. Thus, the opening time is varied only within the first millisecond. Three opening times, i.e. 0, 530, and 800 $\mu$s are applied to unfold the impact of opening time on the high-current anode modes in case of AC 50 Hz. The contact material CuCr7525 and electrodes with diameter of 10 mm are selected. In all cases the first high-current phenomenon is anode spot type 1 that appears at a gap length of 3.0-3.2 mm and the instantaneous current varies within 3.10 and 3.30 kA. A small increase in the arc voltage during the formation of anode spot type 1 is captured. A second high-current phenomenon can be observed over the current half wave depending on the contact opening time. The second high-current phenomenon appears as an anode spot type 2 with an abrupt change in the arc voltage, if the arcing time increases, i.e. separation of the electrode at the very beginning of the pulse (see Fig. 4.9c). By increasing the opening time to 530 $\mu$s, the second high-current phenomenon – footpoint – is identified (see Fig. 4.9b). Finally, no second high-current phenomenon is observed at opening time of 800 $\mu$s (see Fig. 4.9a) [92].

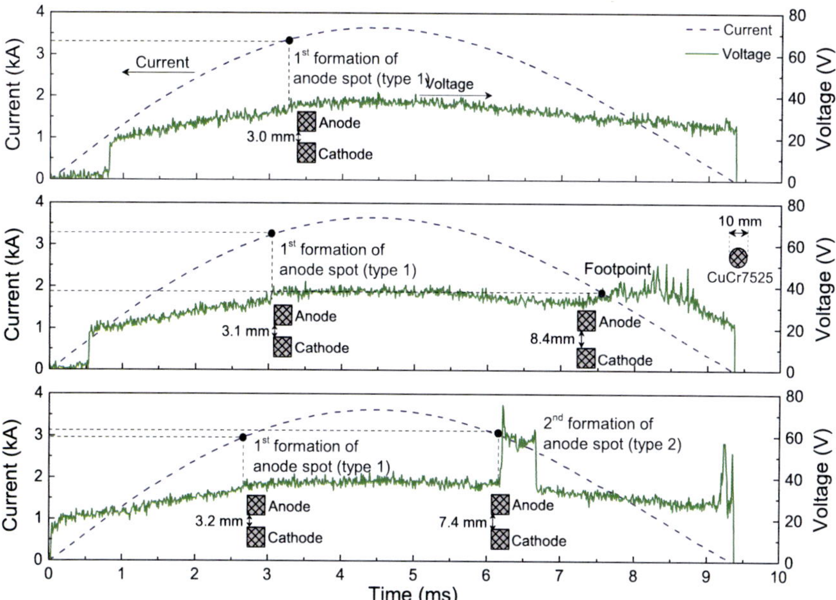


**Fig. 4.9** Voltage and current of the arc including high-current anode phenomena for AC 50 Hz. Different opening times of 800 $\mu$s (a), 530 $\mu$s (b) and 0 $\mu$s (c) are considered. AC 50 Hz with CuCr7525 electrodes having diameter of 10 mm are applied [92].

# 4.3 DC pulses

## 4.3.1 Current, voltage, and high-speed camera images

Fig. 4.10 shows current and voltage waveforms together with approximate time for the formation of anode modes in case of pulsed DC with 10 ms. The two cases show both the appearance of diffuse mode followed by footpoint and anode spot type 1. However, they clearly differ in the appearance of anode spot type 2. Here, the maximum current in both cases reaches the value about 2.7 kA. The electrode opening times are rather similar, i.e. about 3.3 ms and 3.5 ms for case 1 and 2, respectively. The average arc voltage during anode spot type 1 is about 35 V. Nevertheless, an abrupt increase in the arc voltage of nearly 20 V occurs in case 2, indicating the occurrence of an anode spot type 2, in which arcing time is higher and about 200 $\mu$s. As already known from the AC pulses, the arc voltage continuously decreases during anode spot type 2 and finally, after about 3 ms it goes down with an abrupt decrease nearly to the same arc voltage as in case 1.

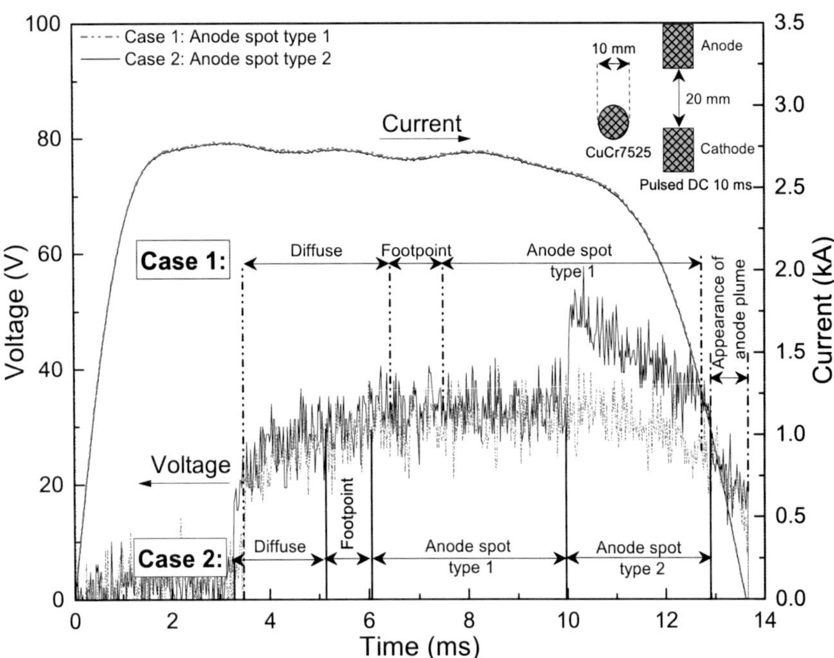


**Fig. 4.10** The current and voltage waveforms together with approximate time instants of high-current modes for the cases 1 and 2, i.e. with and without transition to anode spot type 2 [73].

Fig. 4.11 reflects the images captured by the high speed camera corresponding to the case 1 in Fig. 4.10 during the formation of diffuse, footpoint, and anode spot type 1. At the beginning of the discharge, a diffuse mode can be observed (Fig. 4.11a). A small illumination area appears on the anode during transition to the footpoint at about 6.2 ms (see Fig. 4.11b). After about 7.5 ms, the anode spot type 1 initiates with higher illumination area in comparison to that in footpoint

(Fig. 4.11c). During formation of anode spot type 1 a dark area arises in the inter-electrode gap which is in a good agreement with the results of AC 50 Hz presented in section 4.2.1. Finally, the discharge is transformed back to the diffuse mode (Fig. 4.11d).

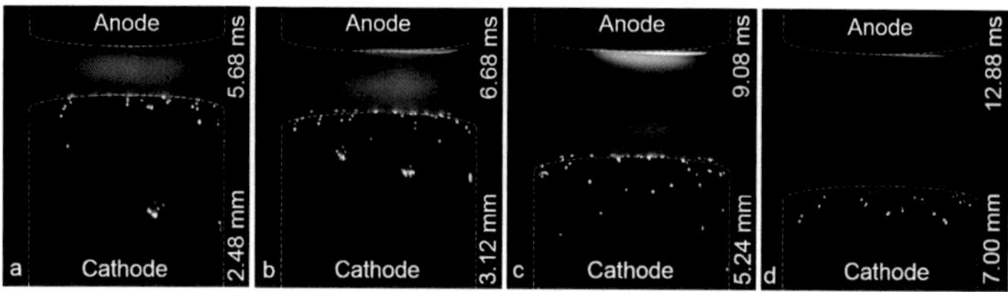


**Fig. 4.11** Video camera images for case 1: a) diffuse, b) footpoint, c) anode spot type 1 and d) diffuse [73].

Fig. 4.12 indicates the formation of each discharge mode including diffuse, footpoint, anode spot type 1, and anode spot type 2 in case 2.

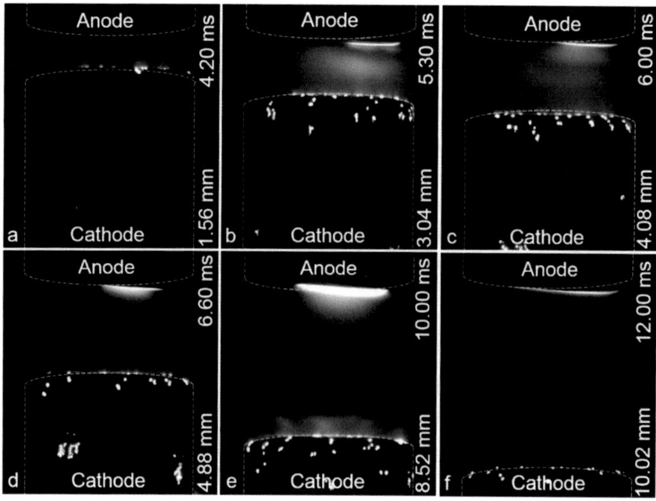


**Fig. 4.12** The formation of different discharge modes for case 2: a) diffuse, b and c) footpoint, d) anode spot type 1, e) anode spot type 2 and f) diffuse [73].

Similar to the case 1, at the onset of the discharge, a diffuse mode can be observed at the anode (Fig. 4.12a), followed by the footpoint until 6.0 ms when anode spot type 1 occurs (Fig. 4.12d). Anode spot type 2 emerges at about 10 ms and gap length of 8.5 mm with considerable larger and brighter illumination areas at both anode and cathode compared to anode spot type 1. It should be mentioned that anode plume is not captured, because the camera is adjusted to maximum intensity during anode spot type 2.

### 4.3.2 Existence diagram

Fig. 4.13a shows the displacement curves for five conducted scenarios with applied pulsed DC 10 ms. It can also be seen when each high-current anode mode appears. The first high-current anode mode, footpoint, is identified at a gap length of about 2.76 mm in case of displacement curve 1. Anode spot type 1 and type 2 are detected at gap length of 4.5 mm and 9.5 mm, respectively. Finally, anode plume mode is formed at current and gap length of about 0.8 kA and 11 mm, respectively. The maximum current is about 2.1 kA. In case of displacement curve 2 the maximum current is about 2.7 kA. Only footpoint and anode spot type 1 appear in this case and even at lower contact gap compared to displacement curve 1 with lower contact speed. The current is interrupted before the electrodes reach the distance of about 8-9 mm that is typically needed for the occurrence of an anode spot type 2. Displacement curve 3 presents the case with the same maximum current of 2.7 kA but with higher contact speed compared to case 2, so both anode spot type 1 and 2 and anode plume appear. Notice, the gap length corresponding to appearance of anode spot modes are lower compared to case 1. Displacement curve 4 shows the conditions in which the footpoint forms at higher gap length no anode spot is observed due to extinction of the arc. Note that a footpoint occurs first in most of the cases presented in Fig. 4.13a; and then the anode spot appears when electrodes reach a certain distance. Therefore, the gap lengths corresponding to anode spots are larger than lengths corresponding to footpoint phenomena. Notice, in case the footpoint appears at higher gap length no anode spot is observed due to extinction of the arc (cf. Fig. 4.14b). Displacement curve 5 shows the conditions in which intense modes appears, in this case electrodes start to open relatively late and the current is higher compared to anode spot modes.

By increasing the number of samples the approximate boundaries between different modes can be determined which is presented by lines in Fig. 4.13b. The threshold currents of footpoint mode, anode spot type 1, anode spot type 2, intense mode, and anode plume are presented by filled triangles, unfilled circles, filled circles, filled diamonds, and half-filled circles, respectively.

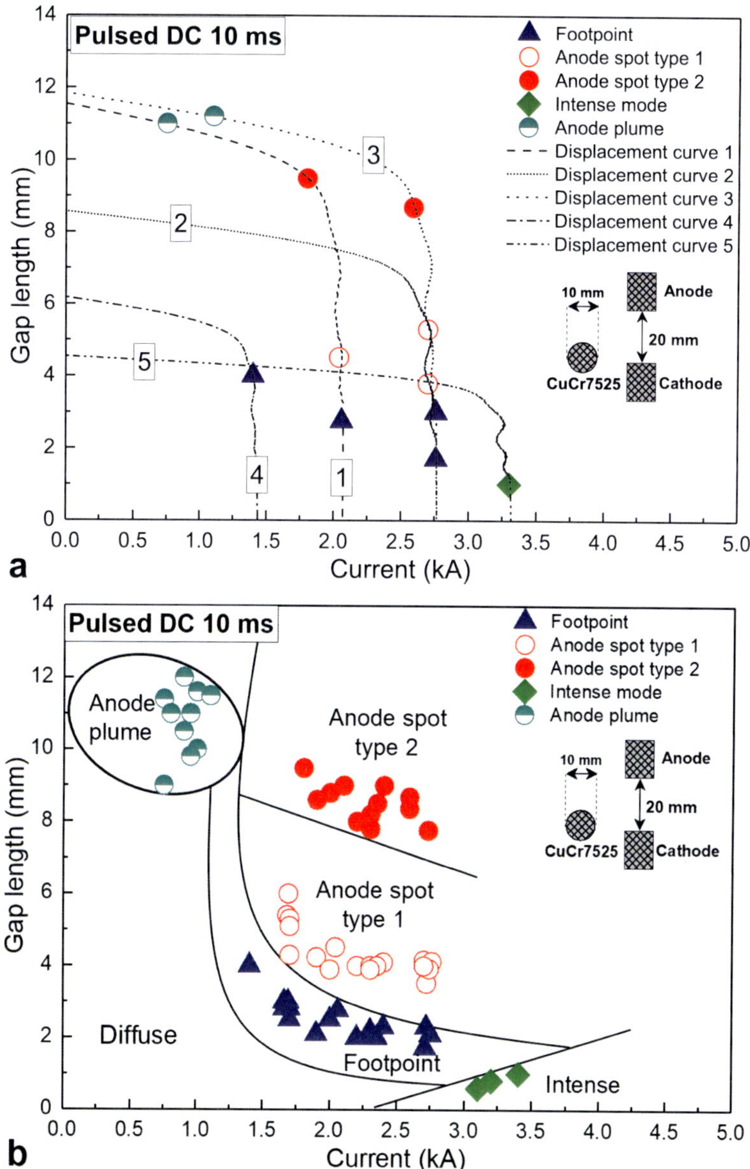


**Fig. 4.13** Existence diagram of high-current anode modes for DC 10 ms together with 5 displacement curves.

## 4.3.3 Pulse length variation

The general behavior of DC pulses concerning high-current discharge modes is investigated applying DC pulses over 5 and 10 ms. In contrast to the above described experiments, the contact material Cu and an electrode diameter of 25 mm are selected. The current and voltage courses together with traveling curves are plotted in Fig. 4.14. The maximum current is about 3 kA in all cases.

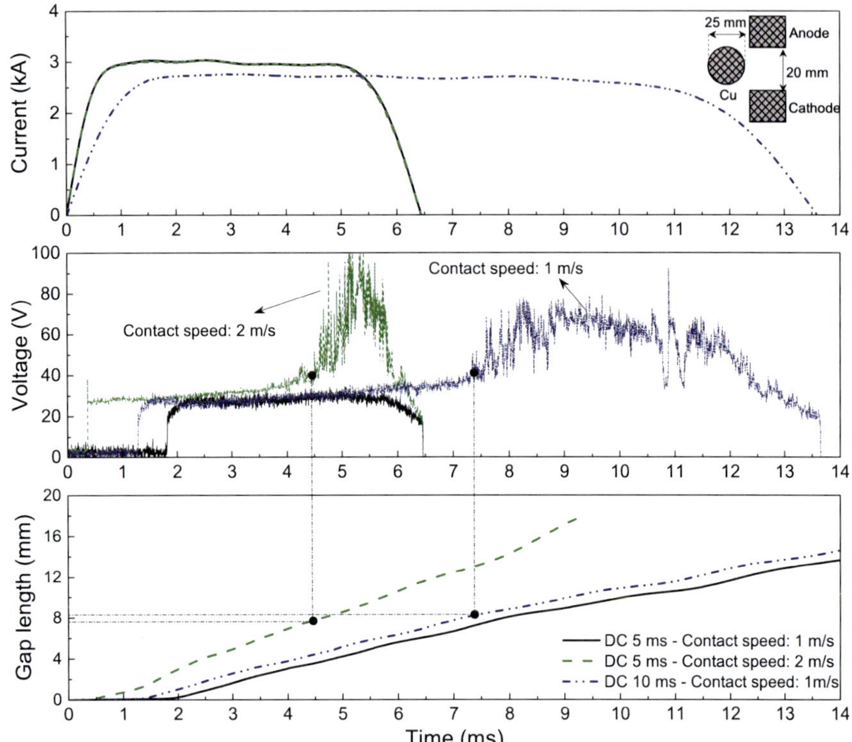


**Fig. 4.14** Current, voltage, and corresponding electrode travel curves for DC pulses of 5 ms and 10 ms. The contact material Cu and the electrode diameter 25 mm are used. Contact opening speed of 1 m/s and 2 m/s is applied [92].

At gap length of about 8 mm, high-current anode modes appear for both DC 5 ms and 10 ms. In the case of DC 5 ms and contact speed of 1 m/s the, arc remains in diffuse mode (cf. Fig. 4.14, black curve). However, by increasing the contact speed to 2 m/s the footpoint mode appears (cf. Fig. 4.14, green curve indicating the increase of the arc voltage and noise). This can be explained by the instability of the anode sheath for larger contact gaps at which the density of the plasma adjacent to the anode is too low [8]. The incident of ion starvation current on the anode will be smaller for the higher contact gap which leads to formation of footpoint in these cases [9].

The formation of high-current anode phenomena corresponding to Fig. 4.14 is recorded by high speed camera and disclosed in Fig. 4.15. The footpoint appears at gap length of 8.5 mm in case of DC 10 ms and contact speed of 1 m/s (cf. the blue curve with increase of arc voltage and noise in Fig. 4.14 and Fig. 4.15c).

In case of long pulsed DC, even by the low contact opening speed the required gap length for the formation of high-current anode mode can be reached [9, 73]. Nevertheless, in such a case the opening time can change the type of anode mode. Although the maximum current of DC 10 ms is lower than for DC 5 ms by 13% (cf. Fig. 4.14a), in DC 10 ms the anode seems to be more active than in DC 5 ms (comparison of the visible emission in Fig. 4.15b and Fig. 4.15c).

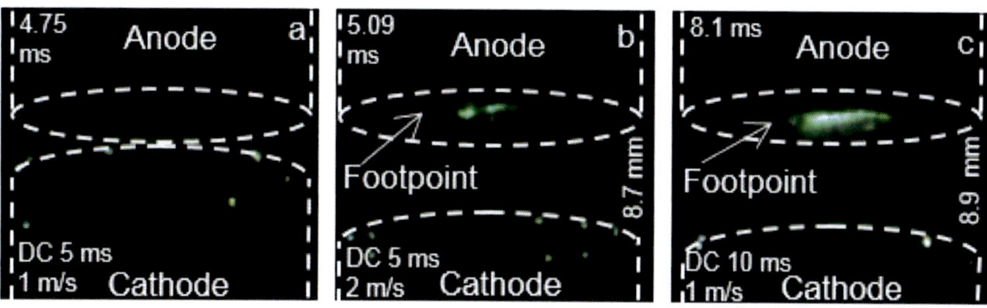


**Fig. 4.15** High speed camera images during the high-current anode phenomena for pulsed DC corresponding to Fig. 4.14. Diffuse mode (a), footpoint (b) and (c) [92].

Considering Fig. 4.14b, significant increase in the mean arc voltage with appreciable fluctuation components are recorded during footpoint mode that indicates remarkable energy loss occurring in the plasma [92].

This increase in arc voltage is due to an increase in the anode drop, which is known to be changed from negative in the diffuse mode to positive in high-current modes [6, 9]. This increase of the anode drop results in a significant rise of the power input to the anode that brings about surface melting and footpoint formation [16].

About 50 shots are taken for pulsed DC 5 ms and 10 ms with varying peak current from 1.5 kA to 5.0 kA to determine the threshold currents of the first high-current anode mode. Dependency of the threshold current on contact opening speed specifically for 1 m/s and 2 m/s is revealed in Fig. 4.16.

For DC 5 ms with opening speed of 1 m/s, the first mode is manifested as intense mode with a threshold current of about 5 kA and gap length of about 2.5 mm, whereas with a contact speed of 2 m/s both footpoint and anode spot are observed at peak current of 3.1 kA and 3.5 kA, respectively, and at gap length of about 7-9 mm. In case of DC 10 ms with both opening speeds, the first high-current mode is found to be footpoint that appears at currents ranging from 2.2 to 2.6 kA and gap length of about 8 mm.

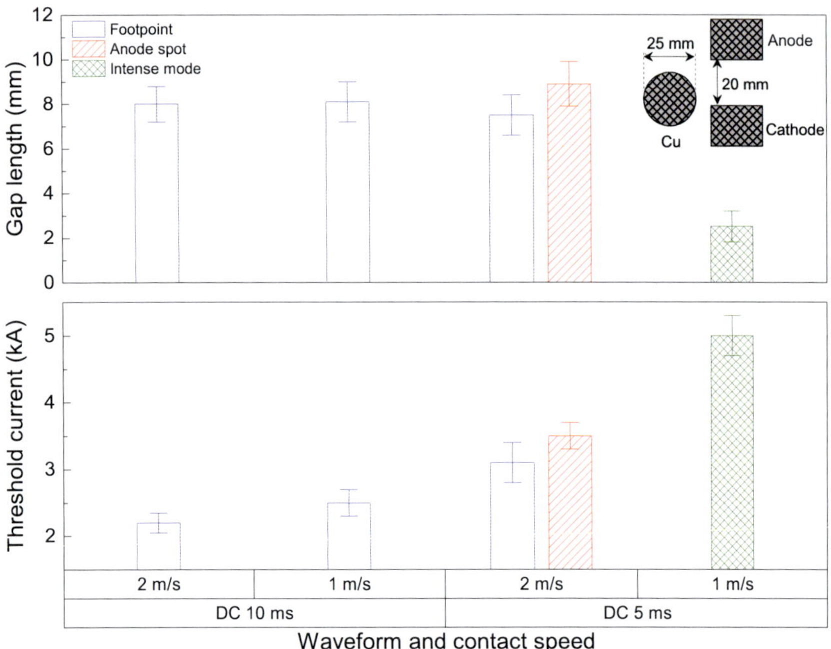


**Fig. 4.16** Threshold current of high-current modes for DC 5 ms and 10 ms pulses. The contact material Cu and the electrode diameter 25 mm are used. Two contact opening speed of 1 m/s and 2 m/s are applied [92].

## 4.4 Contact material and electrode diameter

The threshold currents and gap length in case of DC 5 ms for different CuCr materials and electrode diameters are compared in Fig. 4.18.

By increasing the electrode diameter $D$ and the copper mass fraction, the threshold current $I_{th}$ rises considerably. However, the gap length $g$ in which the high-current anode phenomena appear increases as well, which could be expressed as $I_{th} \propto D/g$ already given in [9].

The enlargement of the electrode diameter alters the current density and consequently the voltage drop near the anode. Thus this influences the power delivered to the anode and the threshold current as well [40].

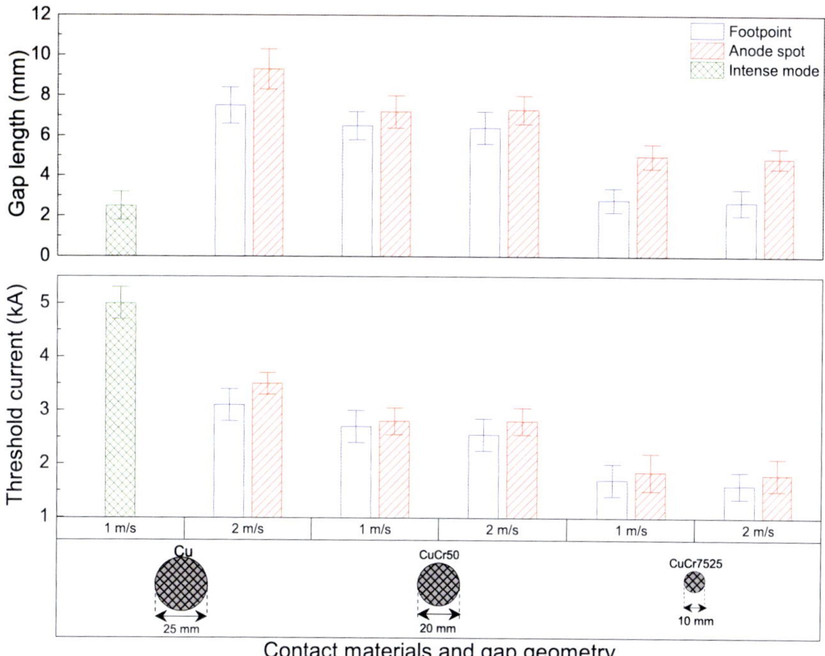


**Fig. 4.17** The impact of the electrodes geometry and material on high-current anode modes. The DC 5 ms is applied and contact opening speed of 1 m/s and 2 m/s is chosen [92].

# Chapter 5

# Optical emission spectroscopy during transition between discharge modes

## 5.1 Introduction

In this chapter, the distribution of atomic and ionic species near the anode, the cathode, and in the inter-electrode gap during discharge modes are investigated by using optical emission spectroscopy. For this purpose the entrance slit of the spectrograph is aligned perpendicular to the anode surface which covers both electrodes. Alternatively, for more quantitative information, e.g. radiating density and electron density during high-current anode modes, the entrance slit is positioned parallel to the anode surface. To get temporal information especially during transition between different high-current modes, video spectroscopy technique is applied [93, 94]. The investigation is performed for both AC 50 Hz and pulsed DC 10 ms with electrodes made out of CuCr7525 and diameter of 10 mm.

## 5.2 Transition between discharge modes

The temporal and spatial distribution of atomic and ionic lines, Cu I, Cu II, and Cu III lines are examined near the anode, the cathode, and in the inter-electrode gap using video spectroscopy. Note that terms I, II, and III stand for atomic, single charge ion and double charge ion species, respectively. The aim is here to investigate the impact of different high-current anode modes on the distribution of atomic and ionic copper lines in order to understand the consequences in terms of arc voltage, electrode loss and switching behavior.

The spectral range of 508 – 517 nm is chosen which contains well resolved spectral lines of different ionization states of Cu, among which spectral lines of Cu I at 510.55 nm, 515.32 nm, Cu II at 508.90 nm, 512.45 nm, and Cu III at 516.89 nm are singled out for this study. Line intensities are obtained by spectral two-dimensional fitting and spectral integration over the line profile. Thus, spatial distributions of spectrally integrated line radiances are obtained for each time. Here, the focus is set on the copper lines because they are more intense than the chromium ones. In fact, the transition probabilities of atomic and ionic chromium lines in range of 508 nm to 517 nm are very low compared to the copper lines [21]. Nevertheless, as presented in [21] the general behavior of atomic and ionic lines seems to be similar for Cu and Cr. To identify the spectral lines, data from several sources including those provided by Striganov, NIST, Zaidel, and Kurucz are used [83–86].

### 5.2.1 AC pulses of 50 Hz

Fig. 5.1 presents the axial intensity distribution of atomic and ionic copper lines during transition from diffuse mode to different high-current anode modes. The approximate position of anode and cathode is also indicated by dashed lines. It should be mentioned that the intensity of each line is normalized with respect to the peak intensity of Cu III during anode spot type 2 (Fig. 5.11).

**Transition from diffuse to footpoint**

The intensity of Cu I near the anode and also in the inter-electrode gap in case of diffuse mode is higher compared to footpoint mode which may be related to sputtering of the anode during the contact separation or effect of cathode spots. The distance between electrodes is only about 1.3 mm in case of diffuse mode which is observed in the early stage of contact separation. In both modes, the relative intensity of atomic line near to the cathode is almost the same.

The intensities of the Cu II line are very low near the anode and has a maximum near the cathode in both diffuse and footpoint modes.

The relative intensity of the Cu III line shows different behavior than Cu I and Cu II with a much broader spatial profile along the discharge axis that increases remarkably by transition to the footpoint mode [94]. Notice that the axial distribution of atomic and ionic lines in diffuse mode presented in Fig. 5.1 shows similarity to those provided by [21] in which the discharge mode is diffuse but with different electrode diameters and maximum currents.

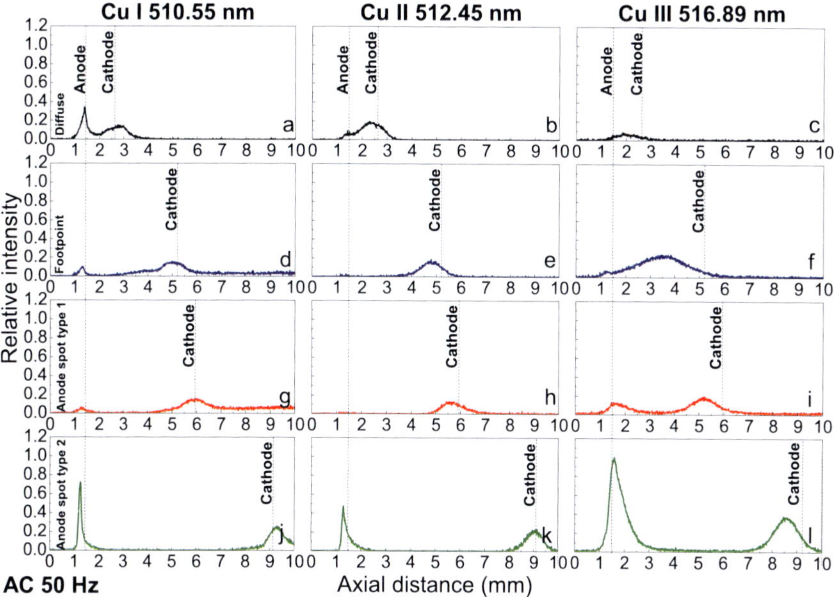


**Fig. 5.1** Axial intensity distribution of Cu I 510.55 nm, Cu II 512.45 nm, and Cu III 516.89 nm during transition from diffuse mode (a-c) to footpoint (d-f), anode spot type 1 (g-i), and anode spot type 2 (j-l) in case of AC 50 Hz.

## Transition from footpoint to anode spot type 1

The intensity profiles of the Cu I line are almost the same in both footpoint and anode spot type 1 with a pronounced increase towards the cathode, but the maximum intensity near the anode is higher in the footpoint mode [94].

The intensities of the Cu II line are very low near the anode and show a broad profile near the cathode in both modes with small changes between footpoint and anode spot mode. During transition from footpoint to anode spot type 1, an abrupt change in intensity appears near the anode, the cathode and also in the inter-electrode gap.

The intensity of Cu III shows only one maximum in the inter-electrode gap in the footpoint mode, whereas local maxima near the anode and the cathode during anode spot type 1. Notice, the dark region seen in Fig. 4.4d confirms the decreased intensity of both atoms and ions between the electrodes in the anode spot type 1.

#### Transition from anode spot type 1 to anode spot type 2

This transition occurs typically at larger gap distances and is therefore accompanied with larger dark areas (i.e. low intensities of Cu I, Cu II and Cu III lines in the inter-electrode gap). Intensities of all species increase considerably near the anode and more moderately near the cathode [94]. However, the intensities of atoms and ions show intensity maxima near the anode which are considerably higher than the intensities near the cathode in the anode spot type 2. Notice that the larger the change of intensity is, the higher the charge number is.

The high speed camera images show that the transition between anode spot type 1 and type 2 occurs very fast confirmed by a jump in voltage and intensity of Cu I-III (cf. Fig. 4.3 and Fig. 4.4). The results agree well with the anode temperature increase presented in [16].

#### Temporal evolution of Cu line intensities in case of AC 50 Hz

Fig. 5.2 represents the relative intensities of atomic and ionic copper lines as function of time at about 1 mm from the anode surface. The two-dimensional curve fitting for each line reveals that the widths of fitted lines at different time steps are almost constant [94]. The approximate time intervals in which the diffuse mode and the different high-current anode modes appear are distinguished. The five lines Cu I 510.55 nm, Cu I 515.32 nm, Cu II 508.90 nm, Cu II 512.45 nm, and Cu III 516.89 nm are chosen as examples.

During transition from diffuse mode to footpoint mode, the intensities of Cu I and Cu II decrease slightly, and then they increase during transition to anode spot type 1. The Cu III line intensity increases in average with some oscillations. By transition to the anode spot type 2, the intensities of all lines rise extremely and approach to maxima depending on the charged state: At first the maximum is reached for the Cu III line, successively followed by the Cu II and Cu I lines.

Reduction of intensities could be seen again when the current amplitude goes down which is accompanied by a transition back to the anode plume. It is noteworthy to mention that the intensities of Cu I lines remain higher compared to the ionic lines.

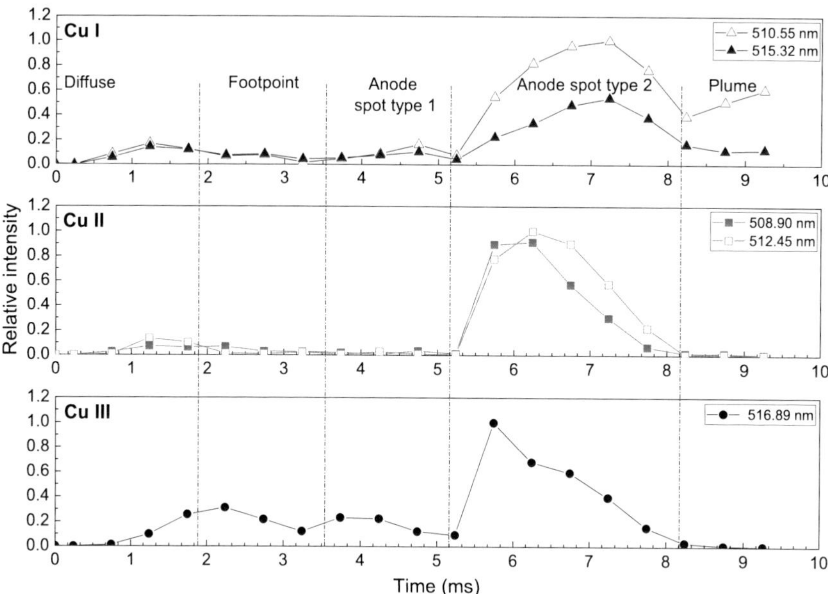


**Fig. 5.2** Temporal intensity of Cu I, Cu II, and Cu III nearby the anode. a) Cu I 510.55 nm and Cu I 515.32 nm, b) Cu II 508.90 nm and Cu II 512.45 nm, and c) Cu III 516.89 nm at about 1 mm from the anode surface. Temporal distributions of different discharge modes are separated by dash lines [94].

## 5.2.2 DC pulses of 10 ms

In this section transition between different discharge modes is explored using pulsed DC 10 ms because of two main reasons. First, the current is almost constant and only the variation of the transferred charge should be considered. The second reason is the increasing interest in the perspectives of a vacuum interrupter for DC switching where high-current anode modes can play a crucial role for the contact erosion and dielectric strength after current zero [73, 93].

Two cases presented in Fig. 4.9 are selected to study the transition between different discharge modes in case of pulsed DC: Case 1 with appearance of diffuse, footpoint and anode spot type 1 and case 2 in which anode spot type 1 is followed by anode spot type 2. These two groups are also termed in the following as case 1 and case 2, respectively (cf. Fig. 4.9, Fig. 4.10, and Fig. 4.11).

**Case 1: High-current modes including anode spot type 1**

Fig. 5.3 reveals the axial intensity distribution of Cu I, Cu II, and Cu III lines during different discharge modes in case 1, i.e. diffuse, footpoint and anode spot type 1 corresponding to the discharge presented in Fig. 4.9 and 4.10. The approximate positions of anode and cathode are illustrated by dashed lines. Notice, the intensity of each species is normalized with respect to the peak intensity of lines of Cu I 510.55 nm during anode spot type 1.

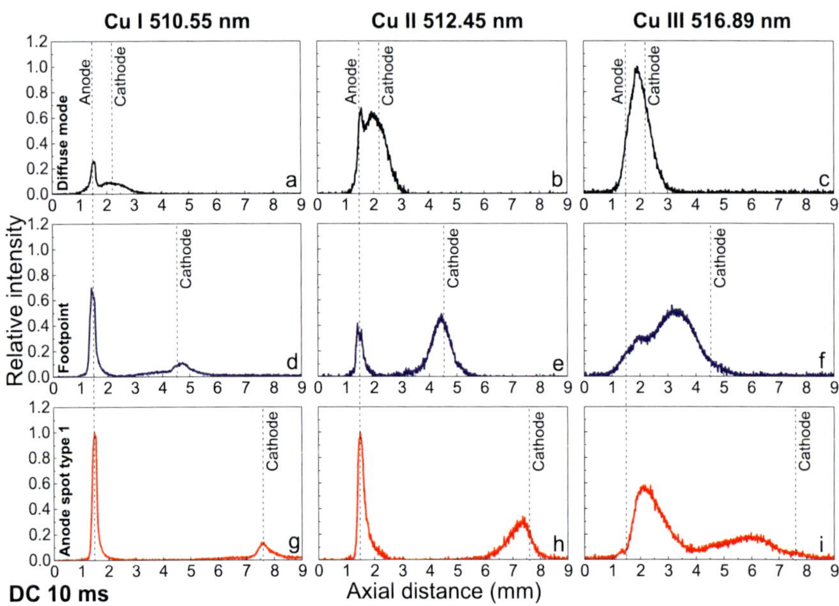


**Fig. 5.3** Axial intensity distribution of Cu I 510.55 nm, Cu II 512.45 nm, and Cu III 516.89 nm during transition from diffuse mode (a-c) to footpoint (d-f) and anode spot (g-i). Maximum current and opening time are 2.7 kA and 3.5 ms, respectively (cf. Fig. 4.9) [73].

The relative intensity of the Cu I line in all discharge modes is higher nearby the anode compared to the cathode. Nevertheless, the intensity becomes higher with every next anode mode near the anode but shows a pronounced decrease towards the cathode. In the inter-electrode gap, the intensity is higher in case of diffuse mode compared to any of the high-current modes, which could be due to shorter gap length in case of diffuse mode.

The relative intensity of the Cu II line during diffuse mode shows profiles with the same intensity close to the anode and the cathode. During footpoint mode, the relative intensity of the Cu II line near the anode and near the cathode is almost the same but with a broader spatial profile close to the cathode compared to the anode. The relative intensity nearby the anode during the anode spot mode is higher than other discharge modes. In contrast, the relative intensity is a little higher during footpoint mode in the vicinity of the cathode.

The relative intensity of the Cu III line manifests a different pattern than Cu I and Cu II (cf. Fig. 5.3c, f, and i). There is a clear change in the distribution of the Cu III line during transition from footpoint to anode spot modes nearby the anode, the cathode and in the inter-electrode gap which is consistent with the results for AC 50 Hz presented in Fig. 5.1.

The intensity of Cu III line has also two maxima in front of anode and cathode during anode spot type 1 but with a much broader spatial profile along the discharge axis than for diffuse and footpoint modes. In fact, from the intensity of Cu III close to the anode it can be concluded that the double charged ions responsible for the Cu III line are generated in front of the anode by collisional processes and not by emitting from the cathode as in the diffuse and in the footpoint mode [73, 93].

**Case 2: Transition to anode spot type 2**

The axial intensity distribution of Cu I-III during different discharge modes in case 2, i.e. diffuse, footpoint, anode spot type 1 and type 2 corresponding to the discharge presented in Fig. 4.9 and 4.11 are presented in Fig. 5.4. The approximate positions of electrodes are also indicated. For comparison, the intensity of each species is normalized based on the intensity of Cu I 510.55 nm during anode spot type 2. Up to the formation of anode spot type 2, the distribution of different species is similar to case 1 presented in Fig. 5.3. By transition to the anode spot type 2, the distribution patterns of Cu I, Cu II, and Cu III lines are similar to anode spot type 1; however, the relative intensities of all species increase during anode spot type 2. It could be also seen that the relative intensity close to the cathode is almost similar to that of anode spot type 1.

The general behavior of the Cu I, Cu II and Cu III lines during formation of different high-current mode in a DC pulse are in a good agreement with the results presented for AC 50 Hz in subsection 5.2.1.

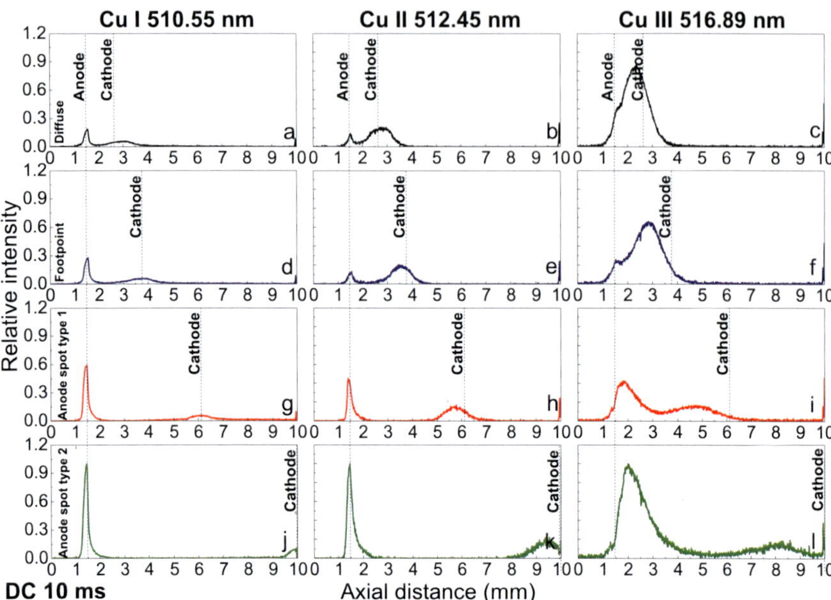


**Fig. 5.4** Axial intensity distribution of atomic and ionic copper lines during diffuse mode (a-c), footpoint (d-f), anode spot type 1 (g-i) and type 2 (j-l). The dashed lines correspond to the according electrode position Maximum current and opening time are 2.7 kA and 3.3 ms, respectively (cf. Fig. 4.9) [73].

## Temporal evolution of atomic and ionic Cu lines in case of DC 10 ms

Fig. 5.5 shows the relative intensities of atomic and ionic copper lines as a function of time at 0.5 mm above the anode surface together with the corresponding voltage waveforms and traveling curves for both cases 1 and 2. The current, which is not shown, is about 2.7 kA for both cases. The relative intensities are calculated based on Lorentz fits of each spectral line at different time steps. Before the formation of anode spot type 2 at about 10 ms, the temporal distribution of Cu I-III follows almost the same pattern. The relative intensities of the Cu I and Cu II lines increase until 10 ms with almost the same slope. During transition to anode spot type 2 in case 2, the relative intensities increase abruptly, corresponding to an abrupt change in the arc voltage in Fig. 5.5 and the integrated light intensity near anode and cathode (cf. Fig. 4.11d and e) [73].

The temporal distribution of the Cu III line is different. It increases first within 1 ms or less and then slowly decreases which can be explained by reconsideration of Fig. 5.3c, f, and i as well as Fig. 5.4c, f, i, and some partially counteracting impacts like electrode distance, rise time of DC pulses as well as electrode heating and material evaporation.

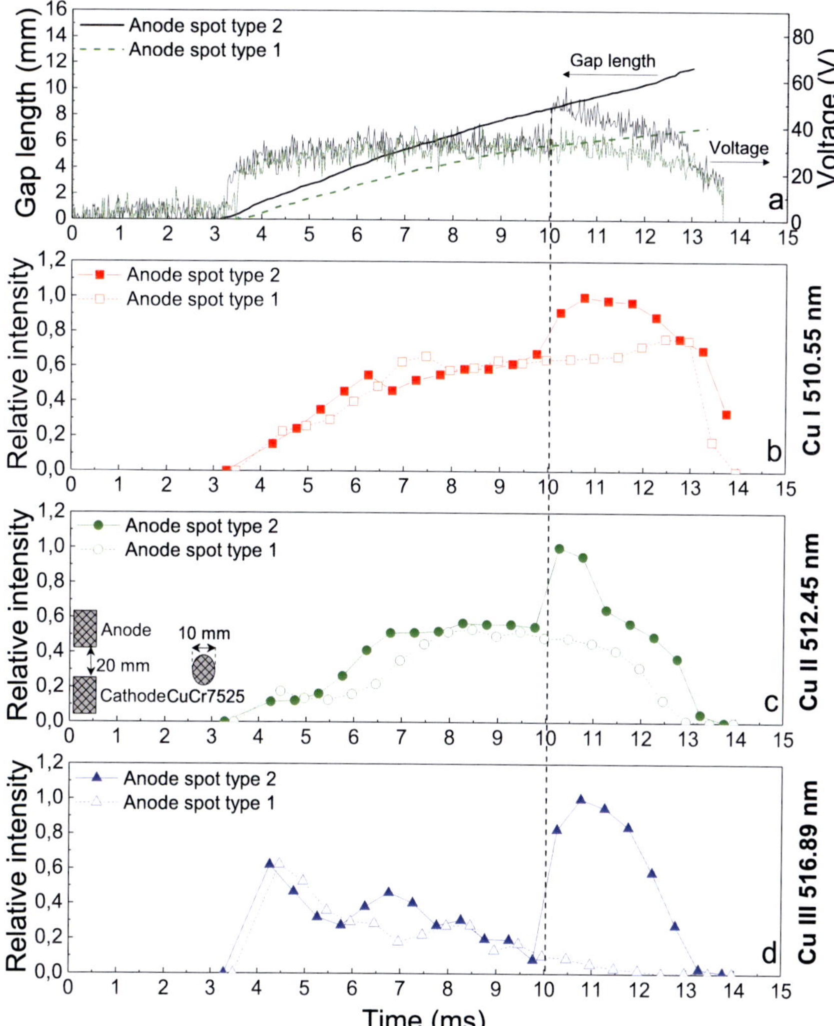


**Fig. 5.5** The relative intensities of Cu I, Cu II, and Cu III lines near the anode as a function of the time for both anode spot types 1 and 2. The dashed line indicates the approximate instant to anode spot type 2 [73].

The initial reduction is due to an impact of the cathode that is the main source of ions at that time and migrates from the anode by time which is confirmed by decrease the intensity of Cu III line near the anode presented in Fig. 5.4c and f. The increase at about 7 ms is related to the formation of the anode spot type 1 that correlates to an increase of particle density of all species nearby the anode due to anode sputtering and evaporation plus ionization of electrons (see Fig. 5.4i). Concerning ionic line emission, a counter effect must occur related to an enhanced evaporation of anode materials since the intensities reach a maximum around 8 ms followed by a slow decrease. It can be either an effect of lower plasma temperature due to cooling by the evaporated material or an effect of charge exchange collisions reducing the average charge number. By decreasing the current at 11 ms and current zero at 13.7 ms, the line intensities drop again which is accompanied by extinction of anode spot. Here, the intensity of the Cu I line remains higher compared to the ionic lines. The relative intensity of the atomic line remains higher even in the course of decreasing current compared to ionic lines that can be explained by the required power input for maintaining the ionization level.

## 5.3 Quantitative analysis of anode spot type 1 and type 2

In this section, anode spot type 1 and type 2 are scrutinized regarding more quantitative information, e.g. radiating density, electron density, etc. The entrance slit of spectrograph is aligned parallel to the anode surface. The results are presented for AC 50 Hz.

**Radiating density during high-current modes**

The local emission coefficients of the Cu I, Cu II, and Cu III lines during the formation of high-current anode modes near the anode can be estimated based on Abel inversion of the measured Cu lines radiation intensities after appropriate fitting with a rotationally symmetric profile and absolute calibration of the intensities [95]. The calculation of radiating densities based on local emission coefficient is described in section 3.3.2. Fig. 5.6 shows the results based on (3.21) for the upper level excited state densities corresponding to the Cu I 510.55 nm, Cu II 505.17 nm, and Cu III 541.84 nm at the distance of 0.1 mm from the anode surface during anode spot type 1 and type 2.

The radiating densities of Cu I at the center of anode spot during anode spot type 1 and 2 are $2.1\times10^{19}\,m^{-3}$ and $6\times10^{19}\,m^{-3}$, respectively. The radiating density of Cu II during anode spot type 2 is about 10 times bigger than anode spot type 1. The density of Cu III shows almost the same distribution during anode spot type 1 and 2. It can be noticed from Fig. 5.5 that the density of Cu II at 4 mm from the anode center position is also 10 times higher than in case of anode spot type 1 at the same position.

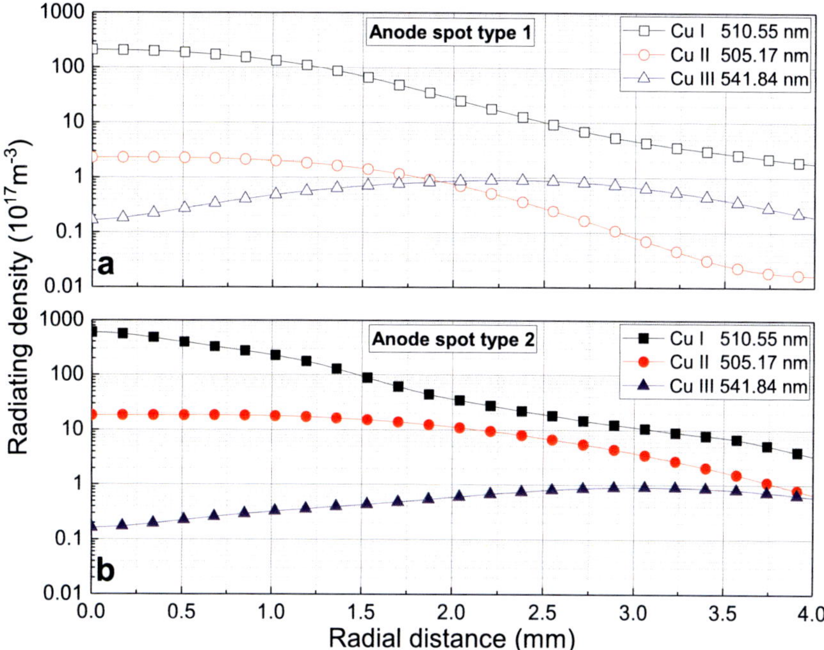


**Fig. 5.6** The radiating density of Cu I 510.55 nm, Cu II 505.17 nm, and Cu III 541.84 nm during anode spot type 1 (a) and type 2 (b) at the distance of 0.1 mm from the anode surface.

Fig. 5.7 illustrates temporal distribution of radiating density of Cu I 510.55 nm at 0.1 mm from the anode surface corresponding to the discharge presented in Fig. 4.1 and Fig. 4.2. During transition from footpoint to the anode spot type 1, the radiating density changes from about $5\times10^{18}$ m$^{-3}$ to about $2\times10^{19}$ m$^{-3}$ and it remains almost constant during anode spot type 1. By transition to anode spot type 2 at about 5.4 ms, the radiating density increases to about $9\times10^{19}$ m$^{-3}$ and after about 1 ms starts to decrease and reaches to about $1\times10^{19}$ m$^{-3}$ at the instant time before formation of anode plume.

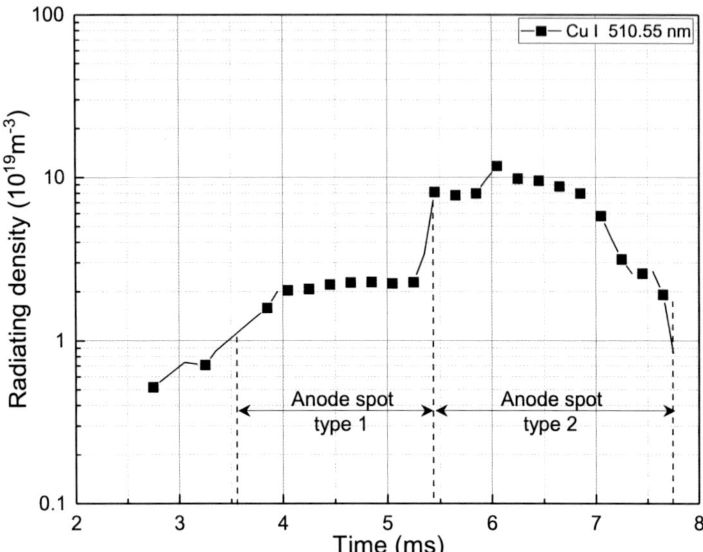


**Fig. 5.7** Temporal distribution of radiating density of Cu I at 510.55 nm corresponding to Fig. 4.3 and Fig. 4.4 at the distance of 0.1 mm from the anode surface.

## Ground state density

It can be assumed for simplicity that excited states of Cu atoms can be estimated in a good approximation by Boltzmann distribution [82]. In addition, the measured Cu I line profiles presented in Fig. 3.13 justify the assumption of low optical thickness of these lines. Under these assumptions, a Boltzmann plot can be used to determine an excitation temperature of the Cu atoms and the ground state density of Cu I [79]. This method is explained in section 3.3.1 and the example is shown in chapter 6 (cf. Fig.6.13).

The Boltzmann plot method has been applied to the excited state densities resulting from measurements during high-current anode modes. Notice, the validity of the Boltzmann distribution is checked using four lines with two different excitation levels. The excitation temperature lies between 8 000 to 11 000 K. Under the rough assumption that the Boltzmann distribution of Cu atom excited states holds up to the ground state according to (3.18), the ground state density of Cu atoms can be estimated. Temporal profile of the ground state density during anode spot type 1 and 2 at the distance of 0.1 mm from the anode surface is depicted in Fig. 5.8.

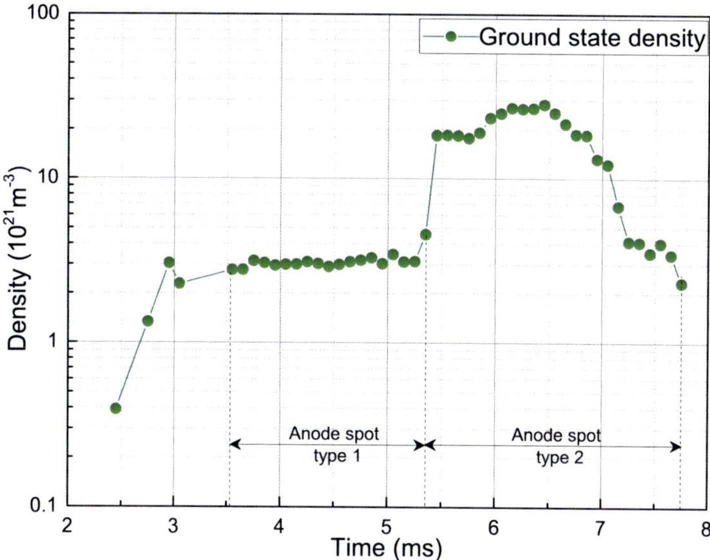


**Fig. 5.8** Temporal distribution of ground state density of Cu I at 510.55 nm at the distance of 0.1 mm from the anode surface.

## Electron density

Stark broadening is applied to determine the electron density during high-current anode modes [20]. The method is described formerly in 3.3.3. The method is employed for different time steps which are acquired by camera. The temporal distribution of electron density during anode spot type 1 and 2 at the distance of 0.1 mm from the anode surface can be seen in Fig. 5.9.

The electron density during anode spot type 1 is about $3\times10^{21}$ m$^{-3}$, while the electron density during anode spot type 2 lies in range the of 10 to $5\times10^{22}$ m$^{-3}$. It should be noted that determination of the electron density during anode spot type 1 is impeded due to technical limitation of the applied method.

## Axial distribution of vapor density and electron density

The results of ground state density, pressure, and electron density during anode spot type 2 at different distances from anode surface are presented in Fig. 5.10. Notice, various shots are considered. The electron density decreases from $7\times10^{22}$ m$^{-3}$ at 0.1 mm to $1\times10^{22}$ m$^{-3}$ at the distance of 1.3 mm.

The ground state density of Cu I increases by factor 2 from 0.1 mm to 0.5 mm. Nevertheless, it decreases by factor about 10 at the distance of 1.2 mm. Meanwhile, the pressure increases by factor 2 from 0.1 mbar to 0.5 mbar and it decreases by about factor 6 at the distance of 1.2 mm.

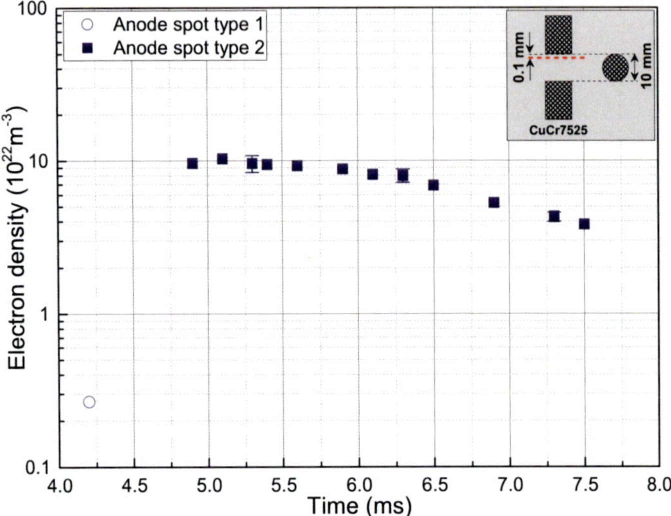


**Fig. 5.9** The temporal distribution of electron density during anode spot type 1 and 2 at the distance of 0.1 mm from the anode surface.

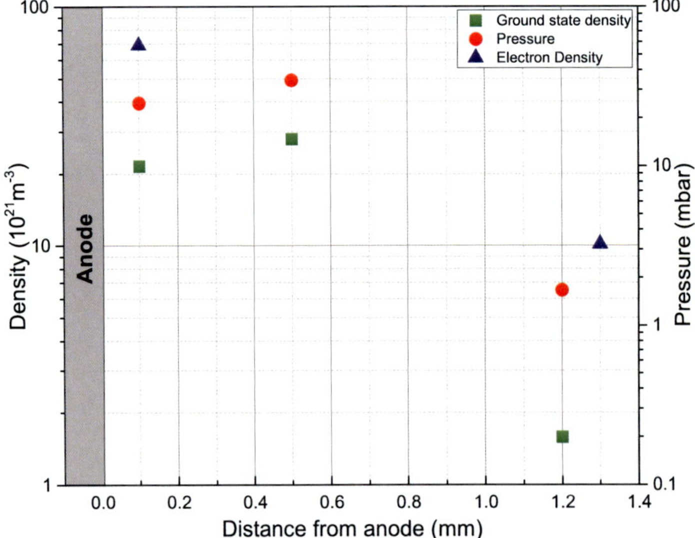


**Fig. 5.10** Ground state density, pressure, and electron density during anode spot type 2 at different distances from anode surface.

## 5.4 Sensitivity analysis

For optical emission spectroscopy presented in this chapter a CMOS camera is used which has a dynamic range of 10 bit. The maximum count number is limited by the exposure time, which is between 10 to 50 $\mu$s to ensure a sufficient temporal resolution. Hence, given 1023 counts in maximum and a noise below 10 counts, the detection limit is expected to be around 1% of full scale intensity. Spectral intensities over atomic or ionic lines depend on correct determination of the underlying background (detector offset, plasma continuum radiation, overlap of other lines) that has to be subtracted. Furthermore, the spectral range of integration must be adjusted properly. Here, the spectra were all modelled by two-dimensional Lorentz fits before integration. Thus, the uncertainty of the calculated relative line intensity is reduced to values below 10%. Note that absolute calibration of spectra is also necessary to conduct quantitative analysis. In this regards an additional error of about 20% should be taken into account for the radiator densities. This is caused on the one hand by the limited accuracy of the calibration curve of the primary normal (tungsten ribbon lamp) and the re-calibration of the applied secondary normal that is used in the lab. On the other hand, the uncertainties resulting from the optical adjustment have to be regarded including apertures, slit width of the spectrometer, aging of the lamp, and transmission of the vacuum windows varying due to deposition.

The applied method of video spectroscopy has two important advantages. Firstly, spatially resolved (two-dimensional) spectra are obtained that allow analysis and comparison of different lines measured in parallel. Secondly, during one shot numerous spectra are acquired under the same experimental conditions including parameters with slight shot-to-shot variation, e.g. electrode opening time, opening speed, surface variations that may cause deviation from rotational symmetry, etc. Thus, concerning most of the results presented in this chapter the shot-to-shot variation can be excluded. Nevertheless, for general description and quantitative analysis, comparison of different shots is needed. Here, quite stable formation of anode spot type 1 and 2 was observed. A maximum uncertainty of the intensity variation of 5% is estimated for the intense lines. However, weaker lines will have higher relative uncertainties. A good example to illustrate the shot toshot variation is shown in Fig. 5.5. Although the cases are different, the results at the beginning of the discharge, i.e. until anode spot type 2 occours, are found to be rather comparable.

Regarding the axial distribution, the widths of all spatial profiles must be in parts attributed to a spatial blurring of the optical system. From geometrical consideration, i.e. the aperture and the slit width of the spectrometer and the pixel size of the detector, it is not expected to measure spatial profiles with a lateral resolution better than 100 $\mu$m, which is close to the minimum widths observed in the experiments (about 200 $\mu$m). In close proximity to the cathode, there is an additional blurring due to electrode movement during acquisition (about 400 $\mu$m in case of 2 m/s opening speed). In some cases, an additional effect can be attributed to radiation from cathode spots located at the side area of the cylindrical cathode. The latter can be controlled with additional observation carried out in parallel by means of a second high-speed video camera.

Regarding side-on measurements, the Abel inversion is determined with respect to rotationally symmetric conditions. Since perfect symmetry is rarely available and the symmetry axis is not always in the center of electrode, some uncertainty of about 20% rises during Abel inversion determination. This mainly relates the deviation at a certain radial position and not absolute values, e.g. in the center.

Regarding ground state density, the validity of Boltzmann distribution is examined using lines with different energy levels. Moreover, non LTE condition could be expected. Here, the error can hardly be estimated without more spectral line intensities of different energy levels and additional simulations.

The electron densities are calculated based on the Stark broadening of atomic lines. The apparatus profile determined from measurements with Hg penrays are rather accurate and the dedicating errors can be neglected. Even more, the broadening itself, i.e. the determination of the Lorentz contribution, is probably not the main source of error. The critical parts seems to be the estimation formula of equation 3.22 that was formulated based on data at a certain temperature and density with relative big uncertainty. Although the temperature might be less critical due to the (rather weak) $T^{1/6}$-dependence, the density factor should be seen critical. Thus, the overall uncertainty can be large, may be even a factor of 3. Nevertheless, this is not valid for relative comparison and often fully acceptable for calculations of absolute densities.

# Chapter 6

# Anode plume

## 6.1 Introduction

The anode spot type 2 is already introduced in chapter 4. The plasma plume is formed after extinction of an anode spot type 2 and before current zero. The plasma plume is observed as an expanding vapor attached to the anode surface and surrounded by a bright shell [41–43]. Based on the investigation of line emission of copper and chromium atoms and singly charged ions using a high-speed camera and interference filters, it is concluded that the inner part of the plume has low radiation intensity and is dominated by atomic line radiation, while a halo of light emitted by ions covers the plume [43].

In this chapter the anode plume formation is investigated systematically using high-current densities. Therefore, AC 100 Hz pulses are applied. Cylindrical electrodes of CuCr7525 with a diameter of 10 mm are used. Video spectroscopy is applied to study the temporal evolution of line intensities of atomic and ionic species including spatial profiles across the anode plume during a single shot.

## 6.2 Anode plume formation

Fig. 6.1 illustrates the existence diagram for the case of AC 100 Hz with electrodes made out of CuCr7525 and diameter of 10 mm. Exemplarily, four displacement curves with opening speed of 1 to 2 m/s and various opening times are presented. The appearance of anode spot type 1, anode spot type 2, intense mode, and anode plume are marked by unfilled circles, filled circles, filled diamonds, and half-filled circles, respectively [82].

Displacement curve 1 indicates the case in which the first high current mode is the intense mode at a current of 4.9 kA followed by the anode spot type 1 that occurs at 6.9 kA but without anode spot type 2. Both types of anode spot occur in case of displacement curve 2 with very early gap opening. The anode spot type 2 is formed at larger electrode distances and even lower currents compared to the anode spot type 1. Displacement curves 3 and 4 reveal the conditions in which intense mode is formed and followed by anode spot type 2. After extinction of anode spot type 2, anode plume emerges.

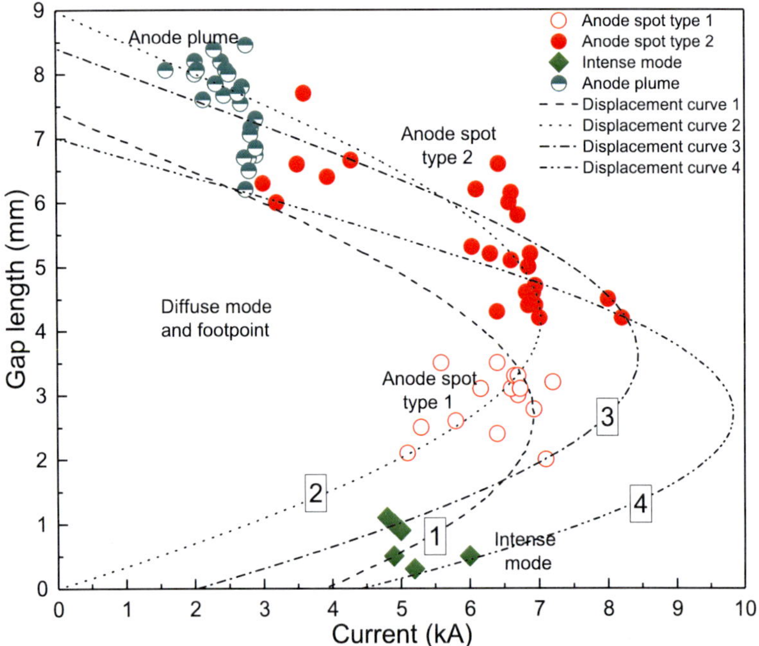


**Fig. 6.1** The existence diagram for AC 100 Hz and CuCr7525 electrodes with the diameter of 10 mm. Four electrode displacement curves are plotted.

Visualized is in Fig. 6.2 an example of current and voltage courses during the formation of the anode spot type 2 according to the displacement curve 2 which is consistent with the results for AC 50 Hz presented in chapter 4. The anode spot type 2 starts at 2.1 ms and a current of about 7 kA. The arc voltage increases during the transition by about 10 V. Anode spot type 2 lasts for about 2 ms. After extinction of anode spot type 2 at about 550 $\mu$s before current zero, i.e. at current of about 2 kA, the anode plume appears (cf. video frames in Fig. 6.3). Notice that the time instants of the corresponding high-speed camera images during the formation of the anode plume are also indicated with the same letters (a-l) as Fig. 6.2.

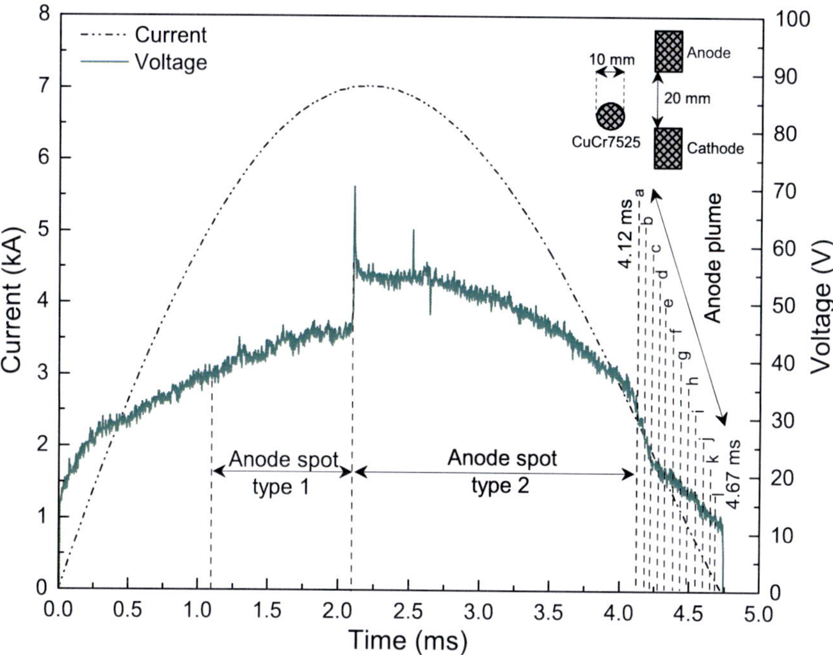


**Fig. 6.2** Current and voltage waveforms together with instant time of the formation of anode spot type 1, type 2, and anode plume [82].

The transition from anode spot type 2 to anode plume occurs within the first three frames of Fig. 6.3. The plume is attached to a bright area at the anode resulting from the previous anode spot. It is characterized by a bright outer shell and an inner region with significantly lower visible emission [43, 82]. The anode plume contracts first (cf. Fig. 6.3c to h) and then it expands (cf. Fig. 6.3i to l) [82]. The intensity decreases by time and is considerably lower than that of the intense mode or anode spot modes. Thus, it might be not observed in video with exposure times that are chosen according to the maximum intensity. Moreover, the input power of arc is decreased considerably by transition to anode plume (cf. current and voltage traces in Fig. 6.2). That leads to decrease of heating of anode surface.

Considering that the source of anode plume is metal vapor [43], a large part of the inter-electrode gap is finally occupied by metal vapor being indicated by the movement and changes of radiating areas in Fig. 6.3i to l. (The existence of metal vapor just practically at current zero is presented in Fig. 5.5b).

The anode plume is a form of appearance of the arc column which is not considered in previous classifications, i.e. diffuse arc, constricted arc, cathode or anode jet arc [96]. Obviously it is strongly related to high-current anode modes as discussed above. Moreover, it can be expected that arc column appearances in general are related to anode modes [9, 96].

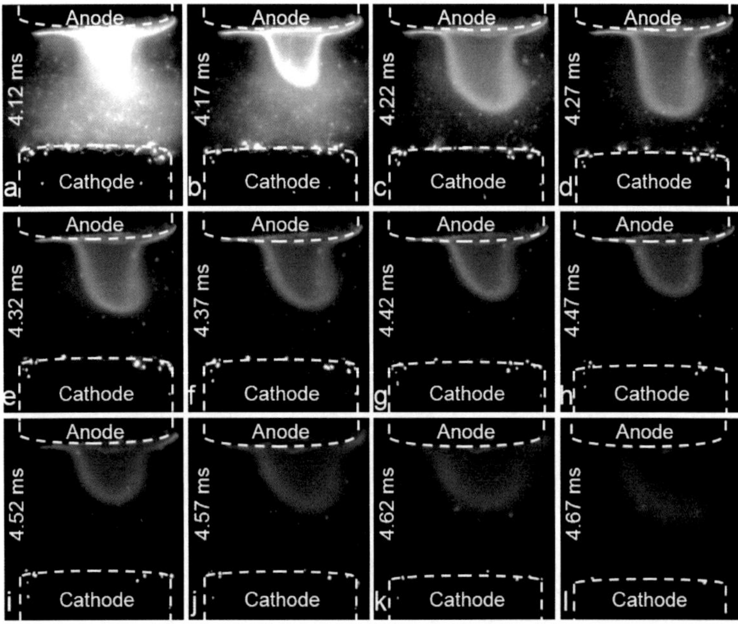


**Fig. 6.3** Formation of anode plume after anode spot type 2, corresponding to the current-voltage waveforms presented in Fig. 6.2. Anode plume is contracted first (c to h) and then is expanded (i to l) [82].

The form and position of anode plume are generally rather stable for several shots under the same experimental conditions. However, it can be changed after some experiments due to high contact erosion and modification of the electrode surfaces [82]. Another example of anode plume after modification of the electrode surface is shown in Fig. 6.4. In this case, two parts of anode plumes appear around the anode center. The displacement curve 3 presented in Fig. 6.1 corresponds to this example. At the time instant of 2.54 ms, the anode spot type 1 is transformed to anode spot type 2 with an abrupt change in the arc voltage. The anode plume is established at about 4.10 ms and after extinction of anode spot type 2 (Fig. 6.4d). Similar to the case presented in Fig. 6.3, it gets slightly constricted (Fig. 6.4d to h), and finally it expands (Fig. 6.4i to l) [82].

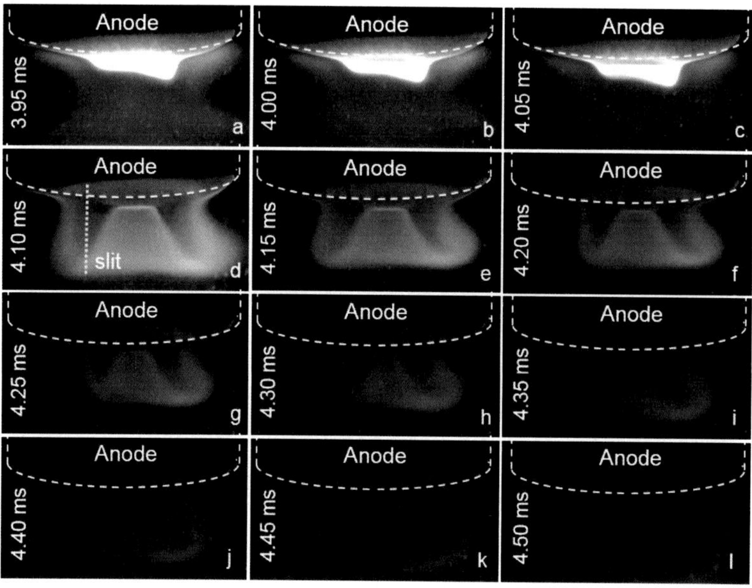


**Fig. 6.4** Formation of an anode plume 400 $\mu$s before current zero corresponding to displacement curve 3. Maximum current equals 8.44 kA [82].

## 6.3 Spectroscopy perpendicular to anode surface

The anode plume presented in Fig. 6.4 is investigated by OES. The spectral range 440 – 580 nm is chosen to study the anode plume. The spectrograph is aligned perpendicular to the anode surface and along the left part of the anode plume as indicated by a dashed line in Fig. 6.4d. The frame rate of 20 000 fps and the exposure time of 48 $\mu$s are selected for video camera. Spectra at any side-on position can be extracted from 2D pattern as shown in Fig. 6.5a and Fig. 6.6a (cf. dashed lines in Fig. 6.5a). Fig. 6.5b and c exhibit spectra near the anode and in the inter-electrode gap before formation of anode plume corresponding to Fig. 6.4c, whereas Fig. 6.6b and c present spectra near the anode and in the inter-electrode gap during formation of anode plume corresponding to Fig. 6.4d.

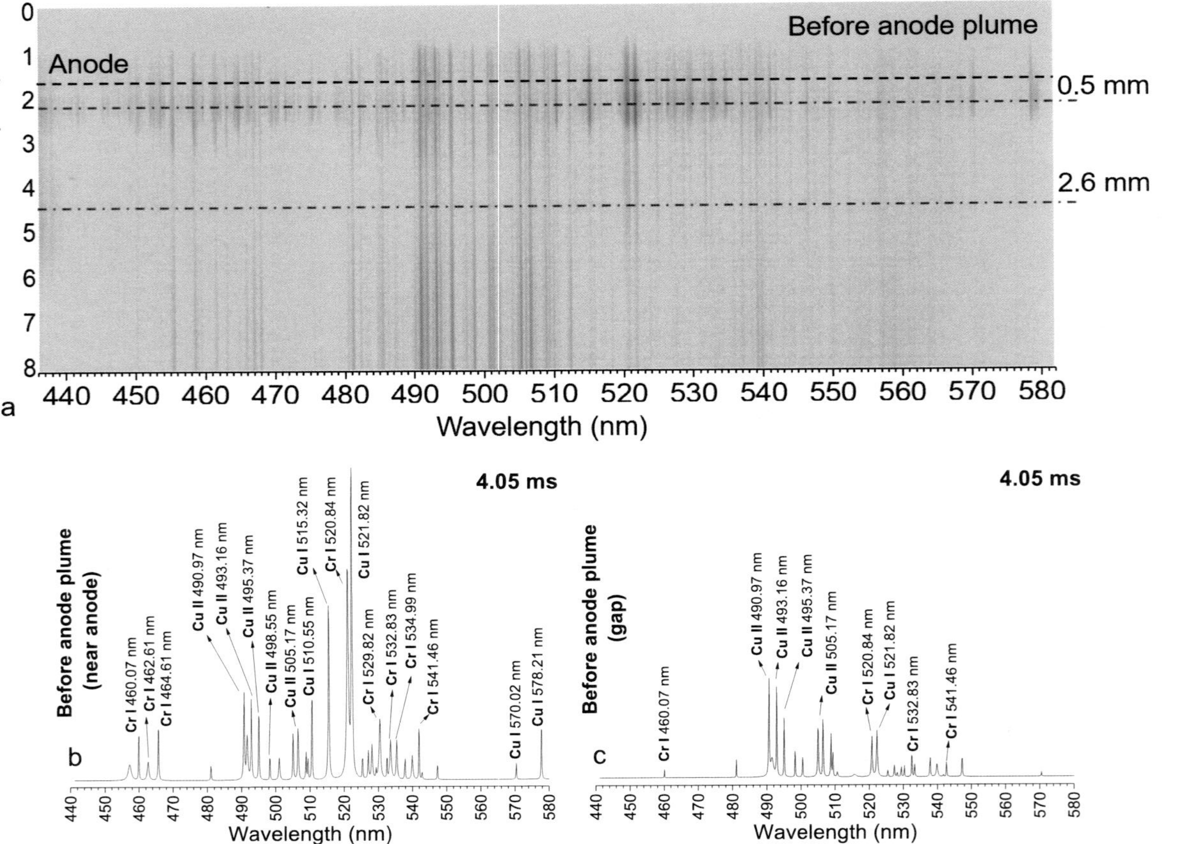


**Fig. 6.5** Spatially resolved spectra perpendicular to the anode surface before the anode plume at 4.05 ms. a) 2D spectrum and corresponding spectral profile for a position b) at 0.5 mm above the anode surface and c) at 2.6 mm (gap region) [82].

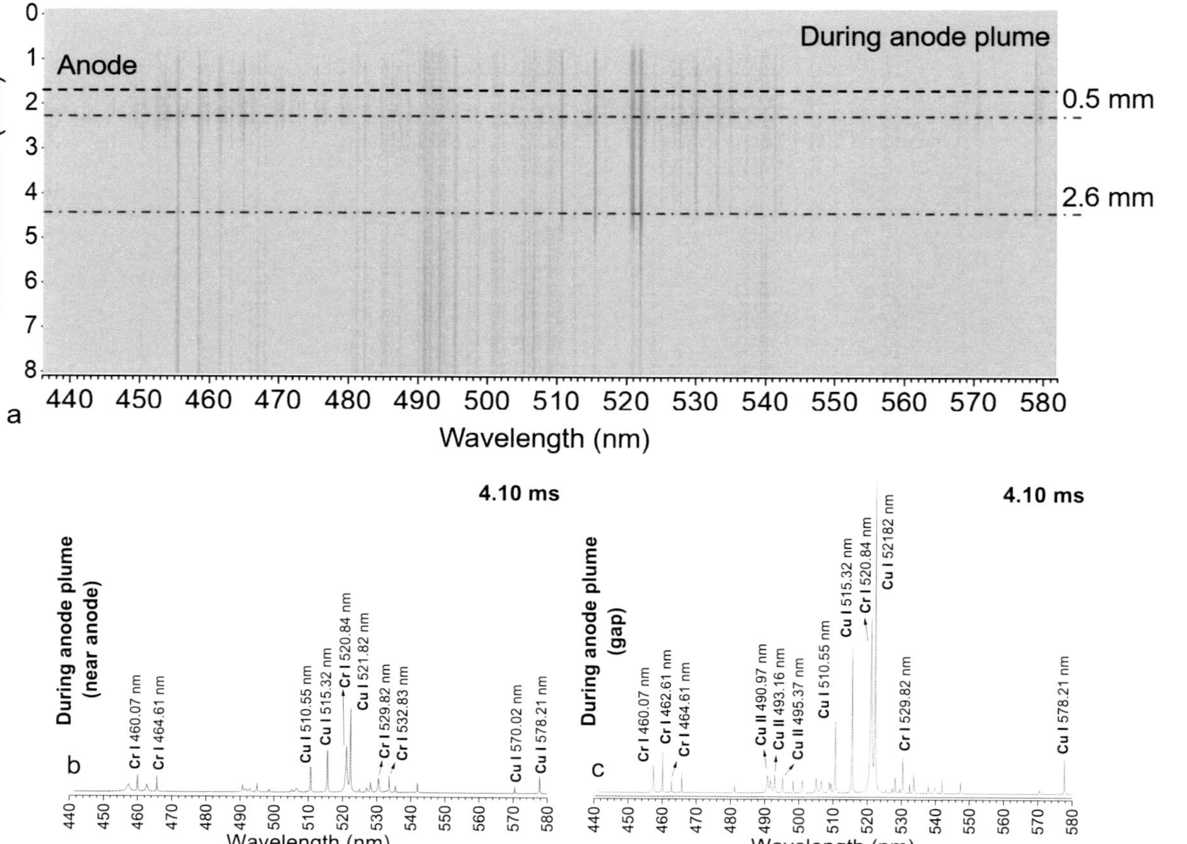


**Fig. 6.6** Spatially resolved spectra perpendicular to the anode surface during the anode plume formation at 4.10 ms. a) 2D spectrum and corresponding spectral profile for a position b) at 0.5 mm above the anode surface and c) at 2.6 mm (gap region) [82].

The intensities of all lines decrease near the anode during the transition to the anode plume (cf. Fig. 6.5b and 6.6b). Simultaneously, the intensities of atomic lines increase and the intensities of ionic lines decrease in the inter-electrode gap (cf. Fig. 6.5c and Fig. 6.6c).

Fig. 6.7 reveals the spatial distributions of line intensities of Cu I 521.82 nm, Cu II 490.97 nm, and Cu III 531.78 nm measured perpendicular to the anode surface before and during anode plume corresponding to Fig. 6.4a, d, and j. Notice, the intensity of each line is normalized regarding the peak intensity of Cu I 521.82 nm at 3.95 ms. The anode position is marked by a dashed line.

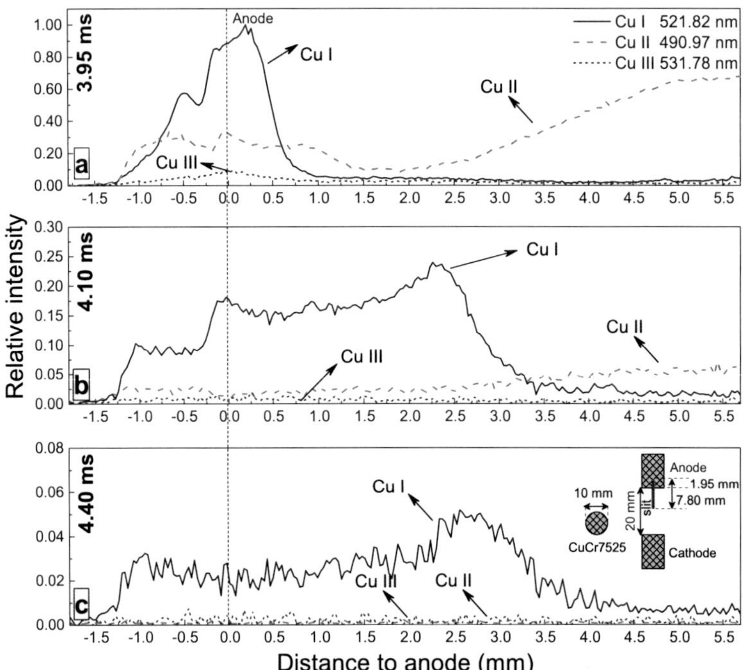


**Fig. 6.7** Spatial distributions perpendicular to the anode surface of Cu I, Cu II, and Cu III at different times prior (a) and during formation of anode plume (b and c) [82].

Before formation of anode plume at 3.95 ms (Fig. 6.7a), the relative intensity of the atomic Cu I line near the anode is higher than that of single ionized Cu II. Moreover, Cu I occupies the area of about 0.7 mm near the anode. In contrast to the region near the anode, the intensity of the Cu II line near the cathode is much higher compared to that of negligible Cu I line. The relative intensity of lines of all kinds are very low measured a few mm away from the anode surface.

Nevertheless, the Cu II line intensity increases towards the cathode which agrees with video spectroscopy results presented in Fig 5.1 and Fig. 5.4.

During transition to the anode plume at 4.10 ms (Fig. 6.7b), the relative intensity of the Cu I line close to the anode surface decreases but it increases in the plume region a couple of mm away from the anode, whereas the Cu II line decreases both near anode and inside the plume, practically to zero. Towards the cathode, the Cu II line intensity still increases. However, the intensity of all species decreases by transition to anode plume. During expansion of the anode plume at 4.40 ms (Fig. 6.7c), only the Cu I line has a noticeable intensity inside the plume. Notice, no Cu III line radiation was detected inside or outside the plume during anode plume phase [82].

The temporal evolution of atomic lines is illustrated in details in Fig. 6.8 for two atomic lines with different upper energy levels at 510.55 nm (3.8 eV) and 521.82 nm (6.2 eV) during formation of the anode plume.

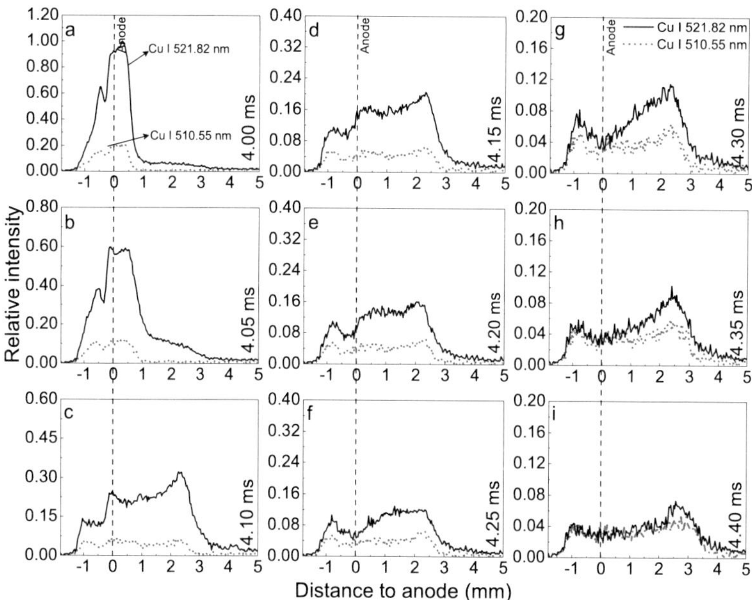


**Fig. 6.8** Temporal evolution of Cu I at 510.55 nm and 521.82 nm during formation of anode plume [82].

The anode position is marked by a dashed line. The intensities of all line profiles are normalized with respect to the peak intensity of the Cu I line at 521.82 nm just before observation of the anode plume at 4.0 ms, i.e. to the same maximum as in Fig. 6.8a. During contraction and expansion of anode plume (cf. Fig. 6.4d to i), the intensities of atomic lines decrease particularly near the anode, while the intensity within the plume is almost constant, and a new maximum at a position about 4 mm occurs at the edge of the plume (cf. Fig. 6.8d to f). But in the course of the final expansion of the anode plume, all intensities rise remarkably toward the cathode (see figures Fig. 6.8g to h).

Even different atomic lines show unequal behavior. The intensity ratio between the Cu I 521.82 nm and Cu I 510.55 nm before and at the beginning of anode plume formation is about of 2-5 (see Fig. 6.8a to d), while it becomes comparable at the end of the anode plume phase. Therefore, considering the higher upper level of the Cu I 521 nm line, it can be concluded that the excitation temperature descends during the expansion of anode plume.

## 6.4 Spectroscopy parallel to the anode surface

To get information on the radial distribution, the spectrograph has been installed horizontally in order to observe the emission parallel to the anode surface as presented by the dashed line in Fig. 6.9. The distance to the anode surface is about 1 mm. Spectra are recorded with 10 000 frames/s and exposure time of 98 $\mu$s. Shown is in Fig. 6.9 the displacement curve 4 with maximum current of 10 kA. Anode plume formation starts at 4.22 ms (Fig. 6.9) which is different compared to those shown in Fig. 6.3 and Fig. 6.4 due to changes of anode surface after numerous shots.

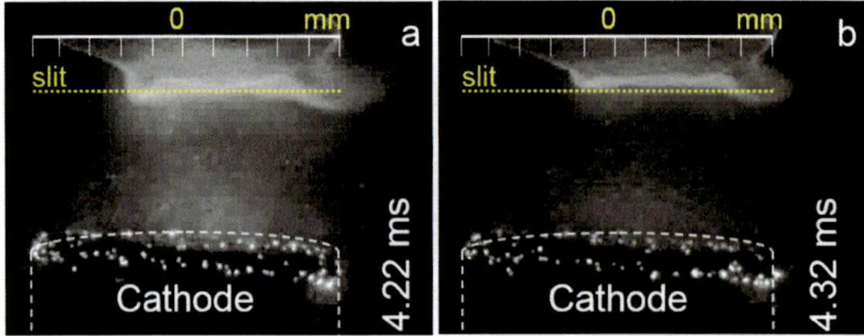


**Fig. 6.9** Formation of an anode plume before current zero corresponding to displacement curve 4 presented in Fig. 6.1. The peak current is about 10 kA [82].

To analyze the differences between atomic and ionic species line intensities are determined, i.e. spectral integration is carried out over selected lines. Fig. 6.10 shows the distribution of relative intensities for Cu I 521.82 nm, Cu II 490.97 nm, and Cu III 531.78 nm parallel to the anode surface corresponding to Fig. 6.9. Considerable differences of radiation intensities from the plume boundary and from inner parts are found. They give reasons to identify them in a qualitative way with the borders and the inner parts as well.

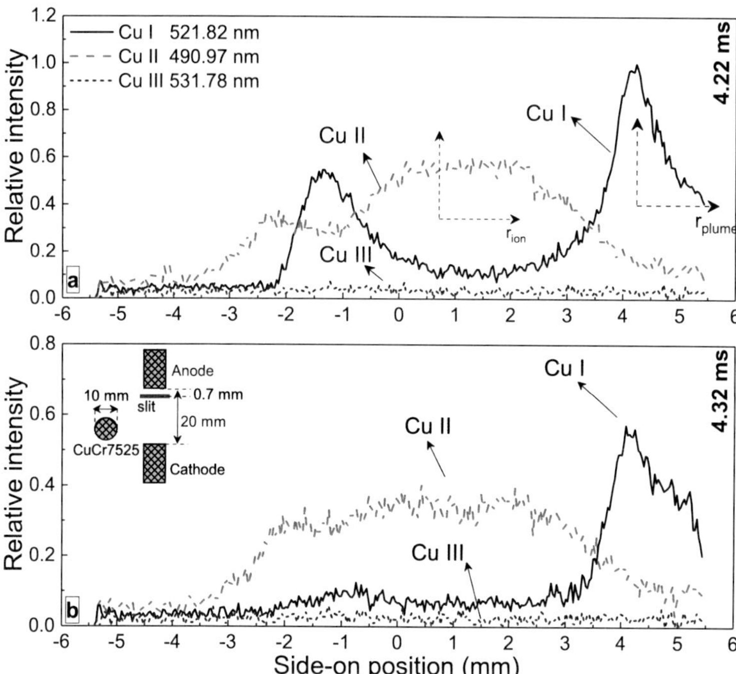


**Fig. 6.10** Distributions of Cu I, Cu II, and Cu III parallel to the anode surface for two time instants corresponding to Fig. 6.9 [82].

The comparison of the anode plume presented in Fig. 6.9a and b with the spectral distributions parallel to the electrode presented in Fig. 6.10a and b confirms that two plume parts are established at both the left and right side of the anode. Moreover, inside the plume is mostly dominated by atomic line radiation, whereas a noticeable line intensity of Cu II is observed in the larger area between the two plume parts and outside, which can be seen at the left hand side in Fig. 6.10b. The intensity of Cu II outside of the plume parts is considerably higher than Cu I which is in a good agreement with [43]. The overall low line intensity from double charged ions proves that a relatively low electron temperature can be expected close to the anode during the formation of anode plume.

## 6.5 Spectra analysis

Only the right side plume part presented in Fig. 6.9a is selected to extract some quantitative results due to inhomogeneity of the anode plume. The radiation intensities are measured side-on only from one viewing angle; rough assumptions for the geometric structure are required to obtain local radiation properties. The assumed plasma structure in front of the anode is illustrated in Fig. 6.11, where a plume part is on one side and is covered by ions.

The plume part on the right side of the anode presented in Fig. 6.9a is assumed to be rotational-symmetric. The axis of the plume part is supposed to be perpendicular to the anode surface at a position corresponding to the maximum atom line radiation intensity of the right part at a distance of about 1 mm from the anode center. Under these rough assumptions, the local emission coefficients of the Cu I lines inside this plume part can be estimated based on Abel inversion of parts of the measured Cu I radiation intensities after appropriate fitting with a rotational-symmetric profile and absolute calibration of the intensities.

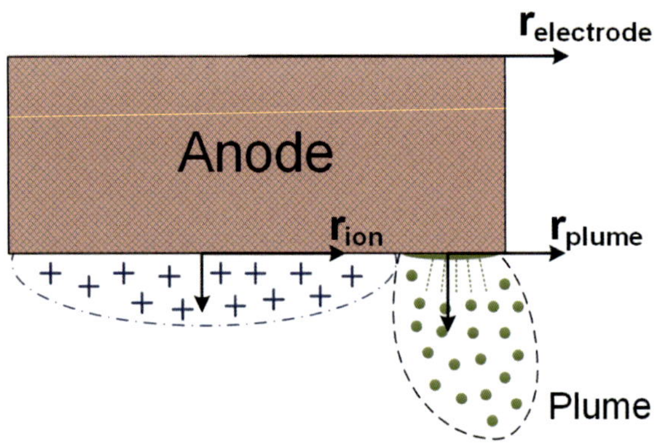


**Fig. 6.11** Schematic representation of plume radius and ion radius [82].

Fig. 6.12a and b present the upper level excited state densities corresponding to four Cu I lines calculated using (3.21) for the right plume. It should be mentioned that meaningful results can be obtained only for the plume part up to the radius of 0.8 mm.

It should be mentioned that the vacuum arc plasma shows considerable deviations from thermal and chemical equilibrium in particular near the electrodes. For instance, the ionization degree at the edge of the shell is not according to

Saha equation. Nevertheless, it can be assumed for simplicity that excited states of Cu atoms are Boltzmann distributed in a good approximation (cf. subsection 3.3.1). In addition, the measured Cu I line profiles (cf. Fig. 6.5 and 6.6) justify the assumption of low optical thickness of these lines.

Under these assumptions, a Boltzmann plot can be applied to determine the excitation temperature $T_{ex}$ of the Cu atoms. Moreover, the right hand term of (3.18) allows also an estimation of the density $n_0$ of the Cu atoms in ground state.

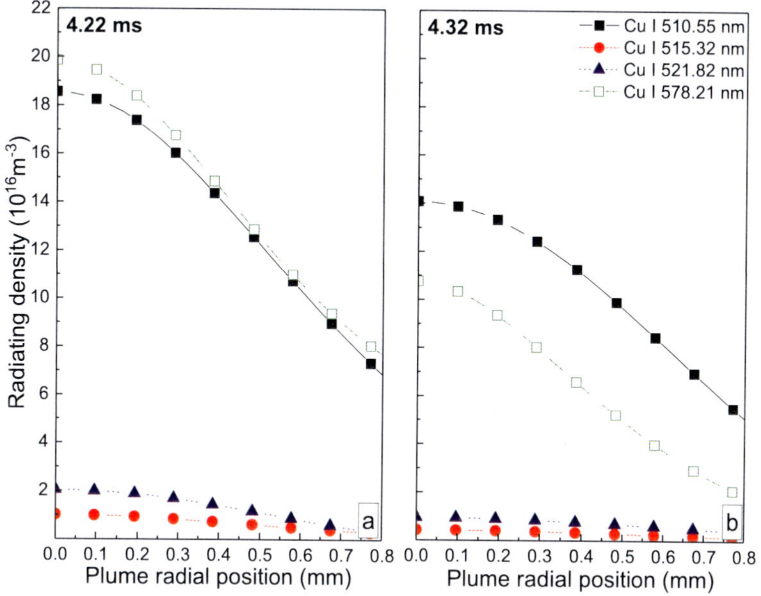


**Fig. 6.12** The radiating density of four Cu I at 510.55 nm, 515.32 nm, 521.82 nm, and 578.21 nm using the equation (3.22) [82].

The validity of the Boltzmann distribution is checked by using four Cu I lines with two different excitation levels presented in Fig. 6.13.

Depicted in Fig. 6.14a is the temporal evolution of excitation temperature based on the Boltzmann plot applied to the exited state densities in the right side of plume for 5 instances before current zero. The temperature varies from 8 500 K to 6 000 K corresponding to times from 4.2 ms to 4.6 ms before current zero. An example of the radial profile of the excitation temperature inside the plume part at 4.3 ms is presented in Fig. 6.14b in which the temperature slightly increases towards the borders of the anode plume.

Under the rough assumption that the Boltzmann distribution of Cu atom excited states holds up to the ground state according to (3.21), the ground state density of Cu atoms can be estimated. The density calculated for different times and positions in the plume part lies in the range of 1-5$\times 10^{19}$ m$^{-3}$. Temporal and spatial profiles of the ground state density are not shown here because of the rough approximation level of the density estimation.

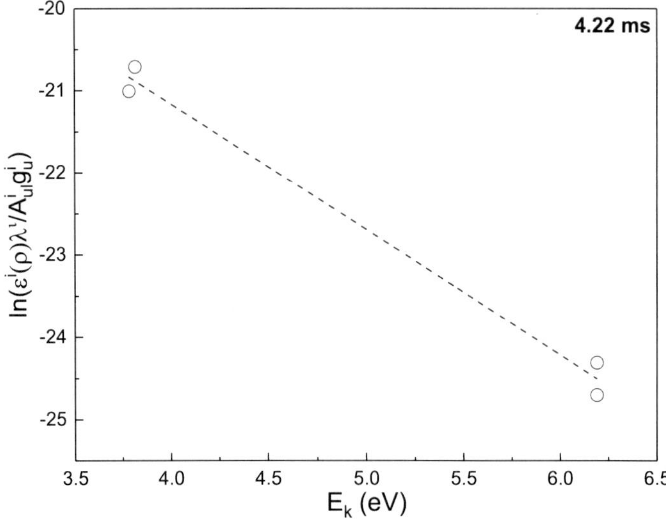


**Fig. 6.13** Example of a Boltzmann plot for the four Cu I lines at the center position of the right side plume part [82].

The radiation intensities of the Cu II line at 490.97 nm can be hardly used for the determination of a local emission coefficient at the positions of the both left and right plume parts which are dominated by atom radiation. A low ionic line emission inside the plume parts is expected. Consequently, the radiation intensity of ions measured side-on may be dominated by a local emission of ions before and behind the plume parts.

A simpler situation regarding the ion radiation near the anode around the center positions of the anode is sketched in Fig. 6.11. Ranges of higher Cu II line intensity as shown in Fig. 6.10 indicate that the Cu II species densities may show a rotational-symmetric structure with an axis near to the anode center position. Under this rough assumption, the radiation intensities of Cu II lines excluding the plume parts are fitted with a rotational-symmetric profile. Abel inversion with respect to an appropriate symmetry axis is applied to calculate local emission coefficients of the Cu II lines.

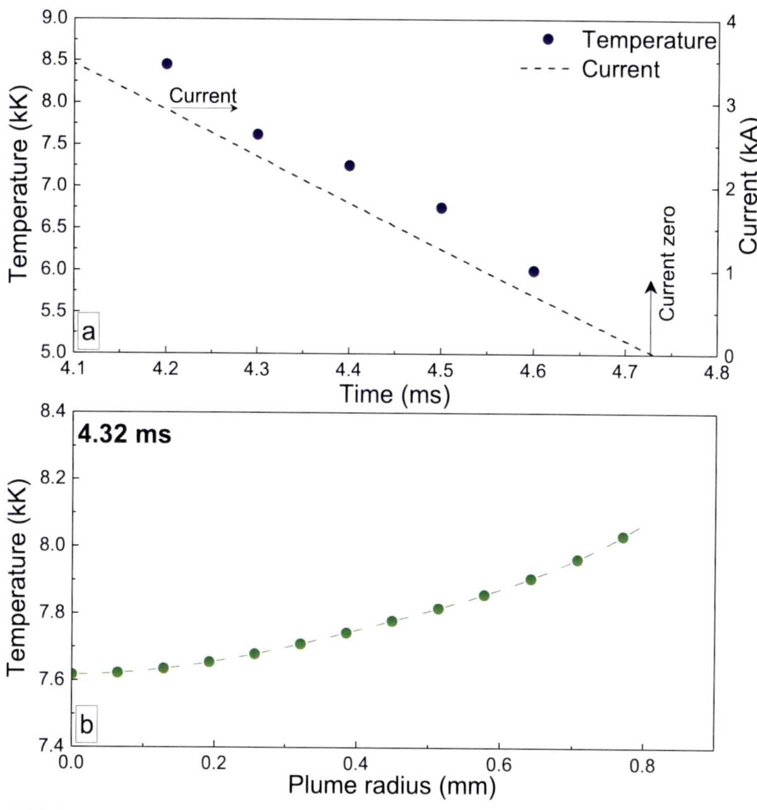


**Fig. 6.14** Temporal distribution of excitation temperature inside the right anode plume (a) and radial profile of temperature inside the plume part at 4.3 ms (b) [82].

The equation (3.21) is used to determine the corresponding radiating density, i.e. the density of the upper excited level of Cu ions with respect to the line. Results for the radial profile of an excited level density are shown in Fig. 6.15 for two time instants.

The estimated excitation temperatures and ground state densities of Cu I s well as of Cu II depend on very rough approximations with regard to the geometric structure of the plasma in the close vicinity of the anode and to equilibrium properties of the plasma.

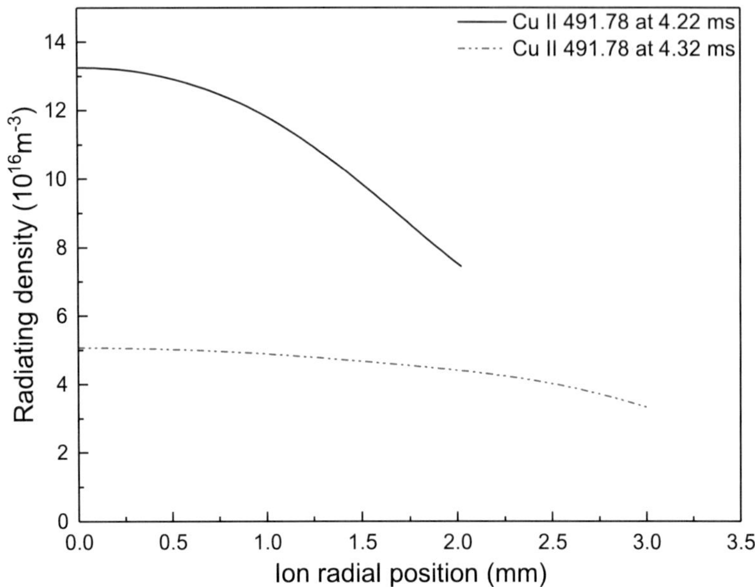


**Fig. 6.15** The radiating density of Cu II at 491.78 nm for two time steps using the equation (3.21) [82].

## 6.6 Sensitivity analysis

A detailed analysis of the shot-to-shot variation of anode plume shows that the anode plume is almost stable under certain conditions, i.e. current amplitude, opening time and speed. Additionally, the distribution of different species parallel and perpendicular to the anode surface is quite stable from shot to shot.

A CMOS camera with a dynamic range of 10 bit is applied in this chapter, too. According to the estimation in section 5.4, a general detection limit is around 1% of full scale intensity

Temperature and density are determined based on Boltzmann plot after absolute calibration and Abel inversion of spectra. About 20% uncertainty due to absolute intensity calibration should be considered, cf. section 5.4. Moreover, applying Abel inversion to such semi rotationally symmetric plasma is rather difficult. At least 100% uncertainty must be taken into account for the Cu I density determination. Due to uncertainties of Abel inversion and deviation from LTE condition, the ground state density could not be determined for ionic lines. The temperatures, however, can be determined with much better accuracy due to the logarithmic factor of the slope in the Boltzmann plot. Typically, an uncertainty of about 10% can be assumed.

# Chapter 7

# Particle density during interruption process

## 7.1 Introduction

In this chapter, the broadband absorption spectroscopy technique [74–77] is employed to determine the Cr I density within first several milliseconds after current zero in case of anode spot type 2 and at 100 $\mu$s after current zero in case of anode spot type 1. Moreover, the Cr I density is measured during high-current anode modes including anode spot type 1 and 2. In this regard, AC 100 Hz as an interrupting current with a peak value of about 7 kA is applied. The electrodes are made of CuCr7525 with diameter of 10 mm; opening speed of 1 to 2 m/s and various opening times are considered.

The spectral range of 423 - 431 nm, which contains three different resonance transitions of Cr I at 425.43 nm, 427.7 nm, and 428.97 nm, is chosen to determine the ground state density of Cu I (cf. Table. 3.6 and Fig. 3.18a). The total absorption within one line is obtained by integration over the spectral line. Accordingly, all the absorption curves are fitted assuming a Lorentz profile.

## 7.2 Cr density determination after current zero with anode plume

2D spectra along the discharge axis at different time instants after current zero is captured using absorption spectroscopy (cf. Fig. 3.18). Cr I density after current zero is calculated regarding (3.26) and (3.29) in case of anode spot type 2 and the subsequent anode plume.

The typical current and voltage waveforms of the considered vacuum arc is shown in Fig. 7.1.

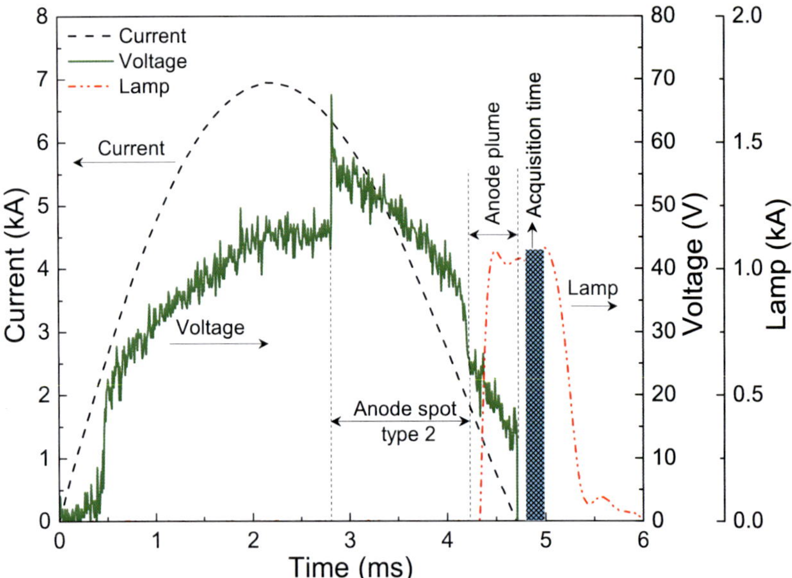


**Fig. 7.1** Current and voltage waveforms of AC 100 Hz arc discharge together with an example for the current pulse of the xenon lamp. The acquisition time of the spectrograph is from 100 $\mu$s to 300 $\mu$ss after current zero (4.8-5.0 ms) [76].

During transition to anode spot type 2 at 2.8 ms, an abrupt change of about 10 V appears in the arc voltage which is in good agreement with the results presented in chapter 4 and 6. A current pulse of the xenon lamp together with acquisition time of spectrograph is also presented in Fig. 7.1. At about 400 $\mu$s before current zero the xenon lamp is triggered. The lamp operates under nearly constant radiation properties. The plateau of the quasi-rectangular pulse has a width of about 1 ms. The spectrograph acquisition is initiated 500 $\mu$s after triggering the lamp, i.e. 100 $\mu$s after CZ at 4.8 ms. The calculated Cr I densities according to (3.26) for 425.43 nm, 427.48 nm, and 428.97 nm lines at 100 Âµs after current zero are $2.41 \times 10^{18}$ m$^{-3}$, $2.31 \times 10^{18}$ m$^{-3}$, and $2.48 \times 10^{18}$ m$^{-3}$, respectively [76].

Accordingly, the temporal evolution of ground state density can be determined by changing the triggering time of spectrograph and xenon lamp. This is illustrated in Fig. 7.2 for a time interval from 0.1 to 3 ms after CZ (begin of acquisition time). The same current is applied and the appearance of anode spot type 2 and anode plume remains stable in all cases. It starts with a Cr density of about $2.4 \times 10^{18}$ m$^{-3}$ and the density decreases by about factor 5 within the first 3 ms. It reduces almost linearly with $0.65 \times 10^{18}$ m$^{-3}$/ms which is consistent with the results presented in [74]. Signal processing for later time instants was not possible due to the detection limit for absorption spectroscopy at the adjusted exposure time of 200 $\mu$s [76].

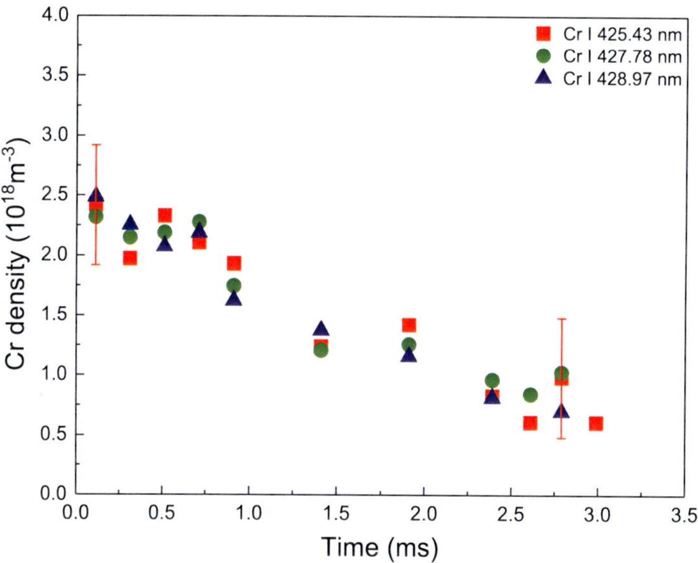


**Fig. 7.2** Cr I density at different time instants after current zero determined from three resonance Cr lines [76].

## 7.3 Cr density after current zero without plume

An abrupt change in the intensity of atomic and ionic lines can be observed during transition to anode spot type 1 and type 2 [73, 94]. Nevertheless, injection of an additional metal vapor to the gap is expected during anode spot type 2 due to the formation of anode plume at the end of this mode. To clarify the influence of anode phenomena on the metal vapor pressure, Cr density at 100 $\mu$s is determined after current zero for the cases without anode plume, i.e. only anode spot type 1 appears. As discussed in chapter 4, formation of the anode spot type 2 depends on opening time or contact opening speed for the same interrupting current [73]. The typical current and voltage traces in case without anode plume are shown in Fig. 7.3. The opening time and speed are 0.7 ms and 1 m/s, respectively. The current is the same as cases with anode spot type 2. However, no abrupt change in voltage could be detected around 3 ms. Meanwhile, the anode is still a source of the metal vapor after the current interruption due to the formation of the anode spot type 1.

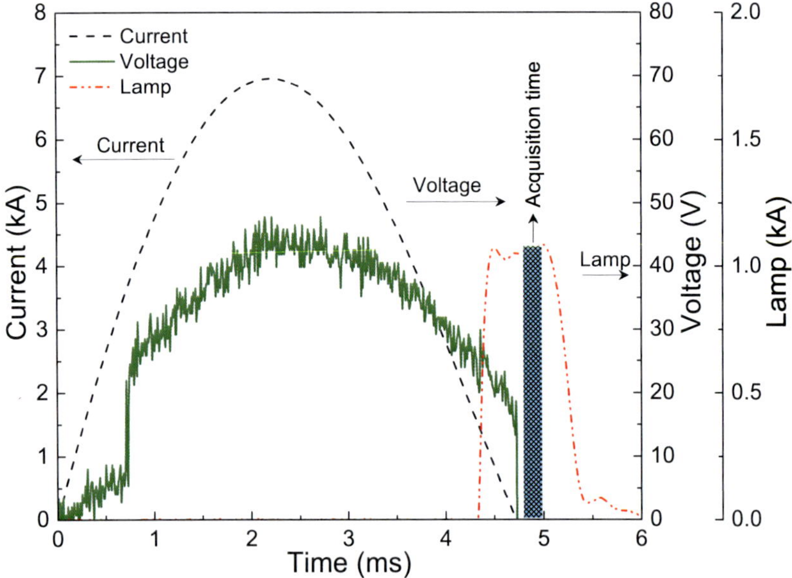


**Fig. 7.3** Typical current and voltage traces in case without anode plume together with current pulse of the Xenon lamp [76].

For control, six independent shots are taken to investigate only anode spot type 1; formation of anode spot type 2 and consecutive anode plume is excluded. According high-speed camera images before current zero are displayed in Fig. 7.4.

The time instants according to CZ in Fig. 7.4 are the same as those for which anode plume is obtained in the example shown in Fig. 7.1. The aspect ratio $D/l$

in case of without anode plume is bigger than case with plume. The gap length is about 8.1 mm and 7.0 mm in case with and without plume at the same time instant, respectively. That means, no anode spot type 2 can be observed if contact opening speed is reduced from 2 m/s to 1 m/s in these experiments. The Cr density in case of without plume at about 100 $\mu$s after CZ equals $1.5\times10^{17}$ m$^{-3}$. Due to weakness of absorption signal, all measurements are performed at the same time, i.e. no acquisition with longer delay after CZ is carried out.

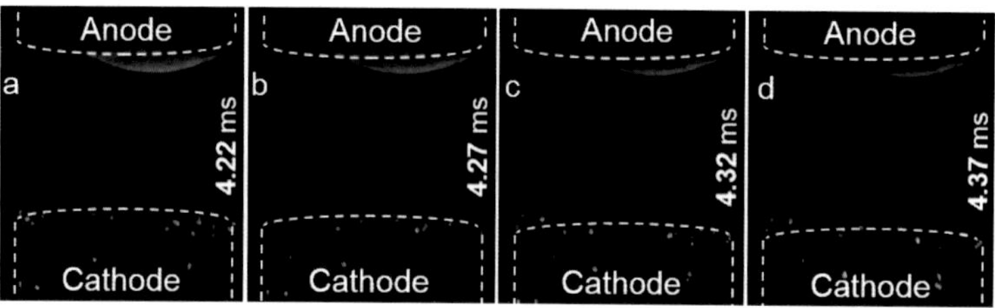


**Fig. 7.4** High-speed camera images before current zero corresponding to Fig. 7.3 in case of anode spot type 1, i.e. without anode spot type 2 [76].

## 7.4 Cr density determination during high-current anode modes

In this section a broadband absorption technique is applied to determine the Cr I density during high-current modes including anode spot type 1 and anode spot type 2. The Cr density during anode plume is also measured and compared with Cr density at the same time instant without anode plume. The same discharge current and electrodes configuration as presented in section 7.3 are implemented.

The current, voltage, and signal of the xenon lamp together with acquisition time of spectrograph for three exemplary shots to study anode spot type 1, type 2, and anode plume are provided in Fig. 7.5. The maximum current is about 7 kA in all cases; thereby different anode modes are achieved by changing contact speed of either 1 m/s (to obtain only anode spot type 1) or 2 m/s (to obtain anode spot type 2 followed by anode plume). The opening time changes due to system jitter [73]. More than 20 shots are taken at different time instants during formation of multiple high-current modes. The calculated Cr densities are shown in Fig. 7.6. It should be noted that semi logarithmic scale is chosen for better visibility since the Cr densities vary by about two orders of magnitude.

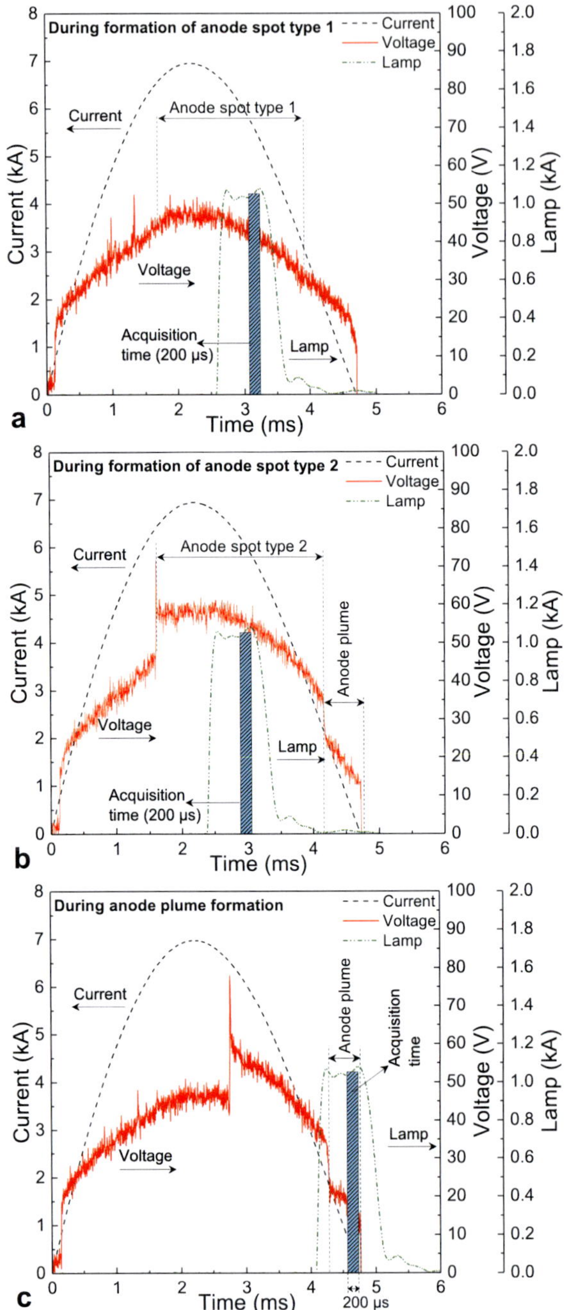


**Fig. 7.5** The current and voltage traces as a function of acquisition time of spectrograph corresponding to anode spot type 1, type 2, and anode plume. The maximum current is about 7 kA.

The density of Cr during anode spot type 1 and type 2 is marked by unfilled circles and filled circles, respectively. The Cr density after extinction of anode spot in case with and without plume corresponding to extinction of anode spot type 2 and type 1 can be seen with filled and unfilled triangles, respectively. The Cr density after current zero is presented here by filled and unfilled diamonds for the cases with and without anode plume, respectively.

The Cr density around 3 ms, i.e. when the current is about 6 to 5 kA, during anode spot type 2 lies in range of $6 - 9 \times 10^{19}\ \mathrm{m}^{-3}$. That is about 5 times higher than Cr density during anode spot type 1 (about $2 \times 10^{19}\ \mathrm{m}^{-3}$). By decreasing the current, the Cr density decreases in both anode spot types. However, it is still much higher during anode spot type 2 than that in anode spot type 1.

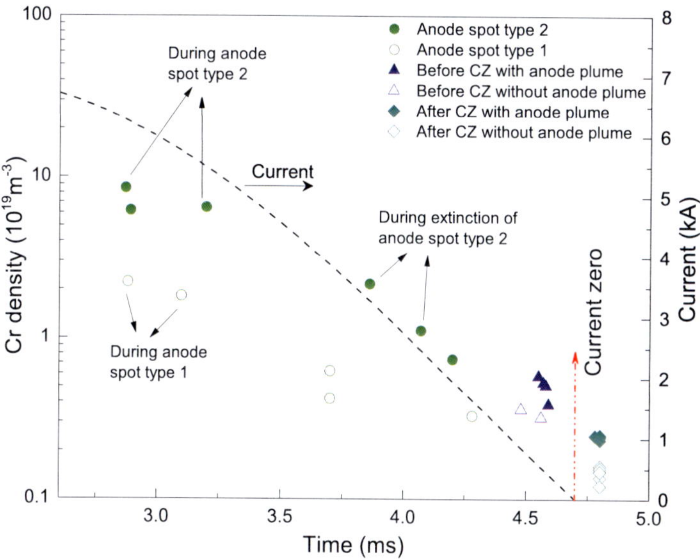


**Fig. 7.6** The Cr density during different high-current anode modes based on the method presented in sections 3.2.4 and 3.3.4. More than 20 shots are performed.

During the formation of anode plume at about 4.5 ms and current around 2 kA, the Cr density reaches the value about $5 \times 10^{18}\ \mathrm{m}^{-3}$. However, at the same time instant but in case without plume (after extinction of anode spot type 1), the Cr density equals $3 \times 10^{18}\ \mathrm{m}^{-3}$. Thus, the ratio of densities for the two cases is decreased to a factor of about two. Immediately after current zero (about 100 $\mu$s), the Cr density for the case with anode plume is still more than $2 \times 10^{18}\ \mathrm{m}^{-3}$

conforming the results presented in Fig. 7.2, whereas it is above $1 \times 10^{18}\,\mathrm{m}^{-3}$ for the case without anode plume in agreement to section 7.3.

## 7.5 Sensitivity analysis

Experimental uncertainties arise from the dynamic range and the noise of the used detector, shot-to-shot variations, and uncertainties in the fit procedure. The intensified CCD camera used for absorption spectroscopy has a dynamic range of 16 bit, which is 65 000 counts in maximum. In the experiments, the maximum count number is limited by the exposure time, which is fixed at 200 $\mu$s to ensure a sufficient temporal resolution. Hence, given 10 000 counts in maximum and a noise of about 100 counts, the detection limit is expected to be around 1% absorption. Extending the spatial region for averaging will improve the signal-to-noise ratio. However, spatial inhomogineities and shot-to-shot variations will limit the benefit of spatial averaging. A more exact quantitative measure for the statistical uncertainty can be obtained from the confidence level of the area under curve when performing Lorentz fits to the absorption profile. It is found that statistical effects contribute to an overall statistical uncertainty of about $\pm 5 \times 10^{17}\,\mathrm{m}^{-3}$, which is about 20% of the density at 100 $\mu$s after CZ (see error bars in Fig. 7.2).

Shot-to-shot variations introduce an additional uncertainty which are exemplary estimated from seven shots 100 $\mu$s after CZ. It reveals that the uncertainty from shot-to-shot variations is between $\pm 0.5 \times 10^{17}\,\mathrm{m}^{-3}$ (with plume) and $\pm 1.5 \times 10^{17}\,\mathrm{m}^{-3}$ (without plume), which is still considerably smaller than the statistical uncertainty discussed above.

Fig. 7.7 shows the appearance of anode plume corresponding to the different time instants after CZ presented in Fig. 7.2.

The position and the size of anode plume remain almost unchanged in different shots at the defined instant of 4.52 ms. However, the shape of the anode plume could be slightly different [82]. Considering the discussed uncertainties, it can be concluded that there is evidence for a temporal decay of Cr I particle densities after CZ for arcs showing anode spot type 2. Even assuming an increasing effective absorption path length for the expanding vapor after CZ, this effect would be rather strong. In cases without anode spot type 2, Cr I particle density is significantly lower and no temporal dependency could be obtained due to the poor signal-to-noise ratio.

Systematic uncertainties would arise from the limited accuracy of atomic constants, e.g. transition probabilities, and the rough estimation of an effective absorption length of constant density. These systematic uncertainties may account for an additional factor of 2. In principle, the uncertainty of the effective absorption length could be in parts reduced by taking radially resolved measurements and performing Abel inversion on absorption signals. The systematic uncertainties do not have any effect on the relative temporal decay of particle densities and the comparison between different discharge conditions, e.g. with and without anode plume, at least as long as the shape and the diameter of the plasma remain unchanged. But it is clear that a decay of density can hardly be measured in cases without anode plume and Cr I density is significantly reduced around CZ due to limited detection sensitivity.

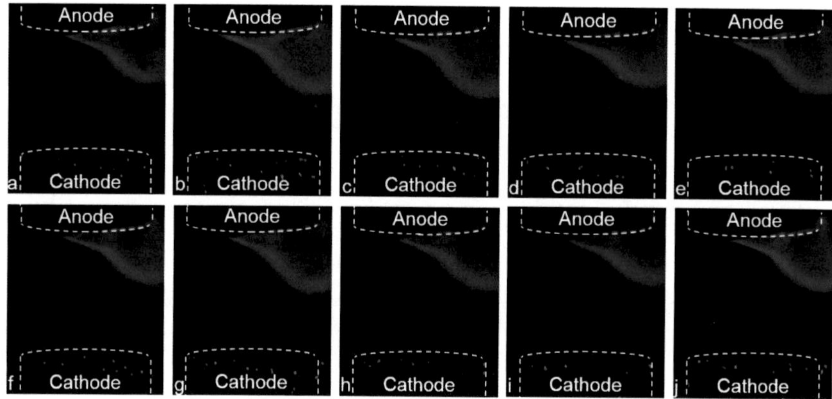


**Fig. 7.7** Appearance of the anode plume for 10 of 11 shots discussed in Fig. 7.2 at different instants of absorption spectroscopy. High-speed images are taken at around 4.52 ms [76].

# Chapter 8

# Arc models

## 8.1 Introduction

In many applications, a black-box arc model is required to predict the interaction between the arc and its surrounding network [97–101]. Notably, to investigate current in HVDC systems, the interaction between the switching device and the electrical grid should be taken into account.

In the simplified electrical arc models, the arc voltage is considered almost constant [52, 57–60], whereas in vacuum arcs during the high-current anode modes the arc voltage may rise up by 50% [73, 82, 94]. This kind of voltage or electrical conductance is caused by the active role of the anode and changes cannot be traced by conventional arc models presented in 2.10.

In this chapter, a new electrical arc model based on the existence diagram for a vacuum arc is proposed, which includes the arc voltage changes during transition between different discharge modes, especially the anode spot type 2. The physical nature of the arc, e.g. the arc temperature, contact geometry and material are considered as implicit parameters in the existence diagrams [9, 73].

This model can be used as black-box model for a wide range of current as well. The model validity is examined by means of AC 50 Hz, AC 100 Hz, and DC 10 ms pulses. The model is realized in EMTP-RV software environment [101].

## 8.2 Proposed vacuum arc model based on existence diagram

The conventional arc models including Mayr, Cassie, Habedank, Schavemaker, Schwarz, Kema, power based model, etc. have been proposed (cf. [57–60]) to simulate and predict current and voltage of the arc. The differences between conventional models are compared in [63]. The results indicate that the differences between models could be majorly traced back in the reignition and distinction of the arc during current zero [102]. However, only the Schavemaker model (expressed in equation 2.4) is singled out as a benchmark for comparison of measured and simulated results. Generally, the arc voltage is almost constant in the conventional arc models [52, 58, 63, 64]. For instance, the average arc voltage in Schavemaker model is denoted by $U_{arc}$. The Cassie arc model (equation 2.3) is typically applied in the high-current regime [51]. In principle, $u_0$ in Cassie arc model estimates the average value of arc voltage. In vacuum arcs, the arc voltage changes during transition from low-current mode to high-current modes within small time intervals. Therefore, the term $u_0$ can be modified to trace the arc voltage during discharge modes.

The transitions from diffuse mode to high-current anode modes can be described by an existence diagram regarding currents and electrode distances [12, 76, 82]. A typical existence diagram of discharge modes for Cu electrodes and AC 50 Hz as interrupting current is presented in Fig. 4.5.

Regarding the new arc model, the arc voltage $u_0$ must be modified by transition to different modes. Therefore, the value of $u_0$ can be written as $[U_h(t) \times (1 - a \times t) + U_0]$, where $U_0$ is considered as fixed parameter for the arc voltage during diffuse mode and depends on contact material. The term $U_h(t) \times (1 - a \times t)$ reflects the arc voltage variation during the high-current modes, where $a$ stands for a calibration factor, i.e. time constant describing voltage increase or decrease during the high-current mode, and $t$ stands for time. $U_h$ is a voltage coefficient related to the transitions between the high-current modes. By determination of the boundaries between diffuse mode and different high-current modes in advance, a data library for the temporal appearance of different high-current modes can be obtained. Based on the time and contact opening speed, which are provided by manufacturer, the approximate discharge mode can be obtained regarding the existence diagram.

The general form of the new arc model for vacuum arcs based on the existence diagram can be presented as

$$\frac{1}{g_e}\frac{dg_e}{dt} = \frac{1}{c_e \times v_e}\left(\frac{ui}{[U_h(t) \times (1 - a_e \times t) + U_0] \times |i|^{b_e}} - 1\right) \tag{8.1}$$

where $v_e$ is a contact speed, $c_e$ is a coefficient of contact opening speed, $a_e$ and $b_e$ are calibration factors. $U_0$ denotes the arc voltage during diffuse mode, and $U_h$ the voltage coefficient related to different high-current modes, which is determined based on the data library regarding the contact speed and existence diagram. Therefore, the model comprises 4 parameters. The contact speed is provided by the manufacture and it is used to find the gap length at corresponding currents which defines the certain point on the existence diagram. $U_h$ in equation (8.1) corresponds to the discharge modes which can provide changes of voltages and electrical conductance during different high-current anode modes. It is noteworthy to mention that the model presented in (8.1) is based on the Cassie but with a variable arc voltage due to high-current anode modes.

To use the new model, it is necessary to establish the dedicated existence diagram (cf. Fig. 4.5 and Fig. 4.15). In Fig. 8.1, a simplified flowchart is presented that shows the general procedure of determination of an existence diagram.

- First, the minimum current for the initiation of anode modes must be experimentally obtained regarding the specific diameter of the electrodes and contact material. As an example, for CuCr7525 with diameter of 10 mm the initiation current lies in the range of 1 kA.

- Secondly, based on the high-speed camera images, the activity of anode and cathode is determined. In this step the characteristics of different high-current anode modes presented in chapter 4 are used to determine the actual type of high-current anode mode.

- In a third step, the current, the gap length and the type of anode modes are saved into the database. Note, that if no activity can be found for the anode, the status is saved as diffuse mode together with the corresponding current and gap length.

- Consequently, the current is increased and the activity of anode will be checked again.

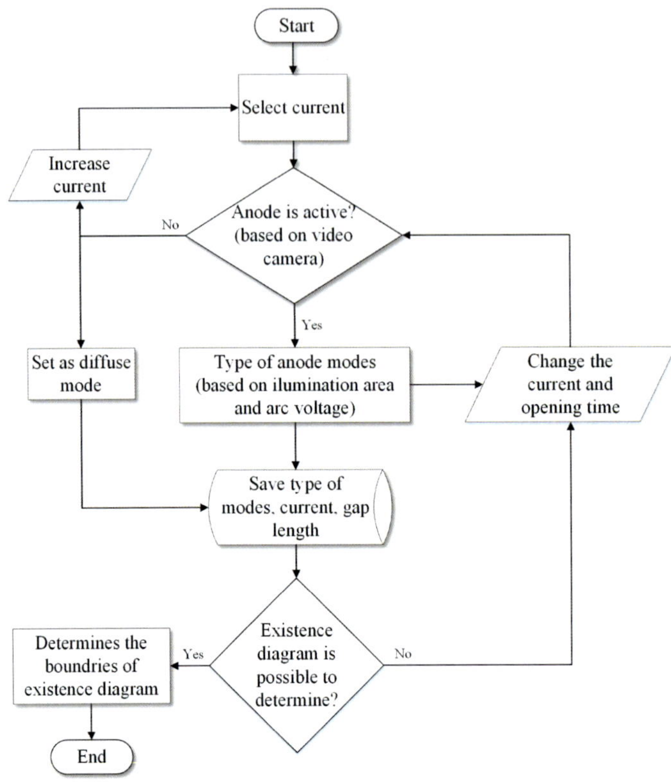


**Fig. 8.1** The simple flowchart that shows the procedure of existence diagram determination.

In case of appearance of high-current anode modes, after saving the data into the database, the current and opening time will be changed to enlarge the number of different modes (cf. Fig. 8.1). In this stage, it is examined whether the number of conducted experiments is sufficient to distinguish the course of the borderlines between the multiple modes in the existence diagram. If not, then the step of changing the current and opening time will be repeated, otherwise the existence diagram can be regarded as solved.

Considering the contact opening speed which is provided by manufacturer, each time corresponds to current and gap length in existence diagram. Different values of $U_h$ in (8.1) are considered for the discharge modes which can be called depending on the time, e.g. $1.1 \times U_0$ and $1.4 \times U_0$ for anode spot type 1 and type 2, respectively. The calibration factors $a$ and $b$ in (8.1) are used to trace either voltage increment or decrement during high-current modes.

For modeling the high-current generator in the EMTP-RV software, capacitors and inductances in parallel or series are applied. The values of $C$ and $L$ elements for each waveform are compiled in Table 3.1 and 3.2. The initial value of voltage of capacitors in EMTP-RV is used as charging voltage.

For simulating the spark gap, a nonlinear controlled resistance is applied which is available in the EMTP-RV nonlinear elements library. For spark gap, Schavemaker arc model is used. By using basic mathematical functions (sum, product, integral, and logical functions), the conductance $G$ as presented in (2.4) is calculated and is used as input for the nonlinear resistance (cf. Fig. 8.2). The same procedure is followed to simulate the vacuum arc, i.e. the nonlinear resistance in EMTP-RV is used and the conductance $G$ for vacuum arc is simulated based on (8.1) and the method described above.

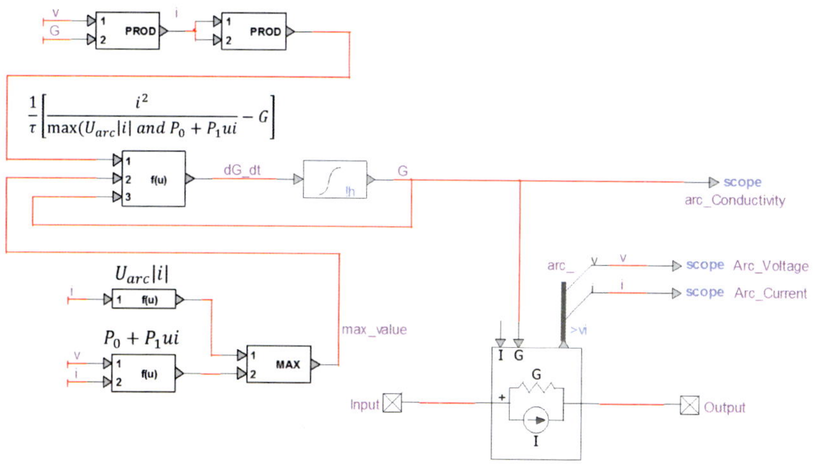


**Fig. 8.2** The simulation of the spark gap in EMTP-RV

# 8.3 Simulation results

## 8.3.1 Conventional electric arc models

All parameters of the models are calculated using the brute-force non-linear least-squares method for optimization including the residual sum of squares (RSS) for the parameter values. This optimization is carried out for the experimental data, and different parameters are obtained for the aforementioned models. Note, that the voltage is considered after opening the contacts only. To evaluate the ability of the model to predict current and voltage traces, the root mean square error (RMSE) and the mean absolute error (MAE) are also calculated.

Fig. 8.3 exhibits the measured and simulated current and voltage waveforms based on the conventional Schavemaker arc model along with the measured values for vacuum arc during high-current anode modes for AC 50 Hz. Due to the dependency of high-current anode modes on gap length [73], two different opening times of 1.5 ms (Fig. 8.3a) and 150 $\mu$s (Fig. 8.3b) are selected to generate different types of high-current anode modes [92]. Notice, the model presented in (2.4) is applied after opening time; therefore, a very small resistance is applied before contact separation. The four parameters $U_{arc}$, $\tau$, $P_0$ and $P_1$ are fitted appropriately.

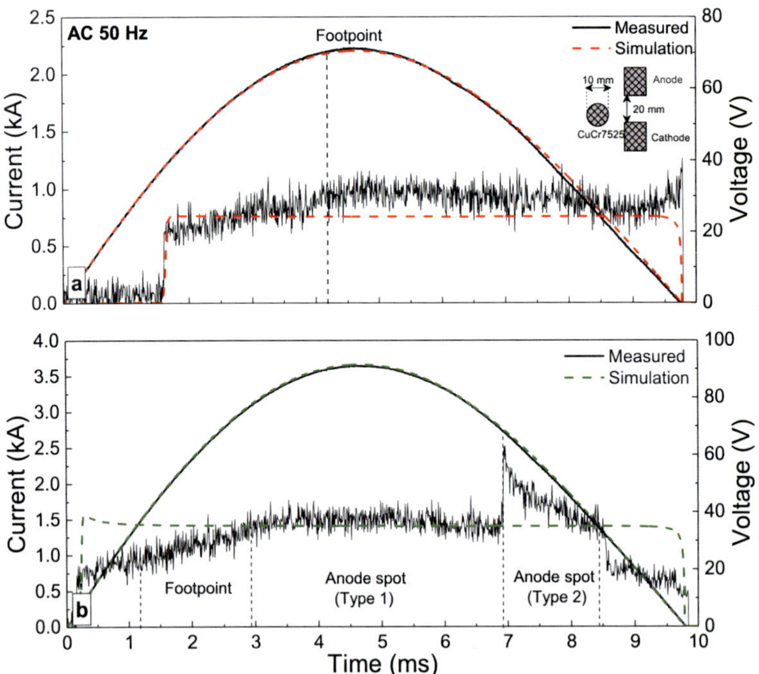


**Fig. 8.3** Measured and simulated current and voltage based on the Schavemaker model.

In Fig. 8.3a, one can see that after contact separation at 1.5 ms a diffuse arc appears; increasing the current at 4.1 ms a footpoint mode appears which is associated with small spots on the anode. During transition from diffuse arc to footpoint mode, the arc voltage increases about 5 V which cannot be traced by the Schavemaker model. However, the current is tracked precisely by the model. In fact, the average arc voltage in the conventional arc models is almost constant and such a change cannot be predicted by those models. Therefore, the value of average arc voltage $U_{arc}$ is fitted based on the average arc voltage during diffuse and footpoint modes.

In Fig. 8.3b, the vacuum arc starts with diffuse mode at 150 $\mu$s. By increasing the current and electrode distance footpoint and anode spot modes appear at 1.5 kA, 3 kA, and 2.8 kA, respectively. The simulated arc voltage is fitted based on the average arc voltage during different discharge modes and applied to $U_{arc}$ in (2.4) resulting in a noticeable difference between measurement and simulation during diffuse mode and anode spot mode. There is a voltage increase of about 20 V during the formation of anode spot type 2 which cannot be detected by conventional models. In fact, the average arc voltage in the conventional arc models is defined by a constant parameter (e.g. $U_{arc}$ in Schavemaker model) that fails to trace the voltage change during high-current modes.

### 8.3.2 Model based on the existence diagram

In this section, the adaptation and use of the new model based on the existence diagram for high-current anode phenomena is illustrated by an example. The existence diagram of high-current anode modes is shown in Fig. 4.5 as function of gap length and threshold current for AC 50 Hz and contact material of CuCr7525 with diameter of 10 mm.

By applying the boundaries between diffuse and high-current anode modes as data library for the model presented in (8.1), the current and the voltage of the vacuum arc can be simulated. Note that anode plume is not considered. The contact opening speed is measured in the lab setup or provided by the manufacturer of a vacuum interrupter and it is used to find the gap length at a corresponding current. $U_h(t)$ in (8.1) corresponds to the different discharge modes which can provide voltages and electrical conductivities during high-current anode modes.

Using the model presented in (8.1) and the appropriate parameter set, the temporal course of voltage and current can be studied. Two examples that are presented with labels case 1 and case 2 are selected to evaluate the new model based on the existence diagram. The estimated parameters for AC 50 Hz regarding case 1 and case 2 as a result of the fitting procedure and the constant parameters model such as $v$ are given in Table 8.1.

The measured and simulated current, voltage, and electrical conductance for a vacuum arc with different discharge modes including diffuse, footpoint, and anode spot modes are presented in Fig. 8.4 and 8.5 based on (8.1) and existence diagram related to the case 1.

**Table 8.1** Estimated and constant parameters of the vacuum arc model.

| Symbol | Quantity | Estimated values | |
|---|---|---|---|
| | | **Case 1** | **Case 2** |
| $c_e$ | Coefficient of contact speed | $6.4 \times 10^{-7}\ \mathrm{s^2\ m^{-1}}$ | |
| $v_e$ | Contact speed | 1.3 m/s | |
| $U_0$ | Arc voltage of diffuse arc | 20 V | |
| $a_e$ | Calibration constant | $92\ \mathrm{s^{-1}}$ | $100\ \mathrm{s^{-1}}$ |
| $b_e$ | Calibration constant | 1.39 | 1.37 |

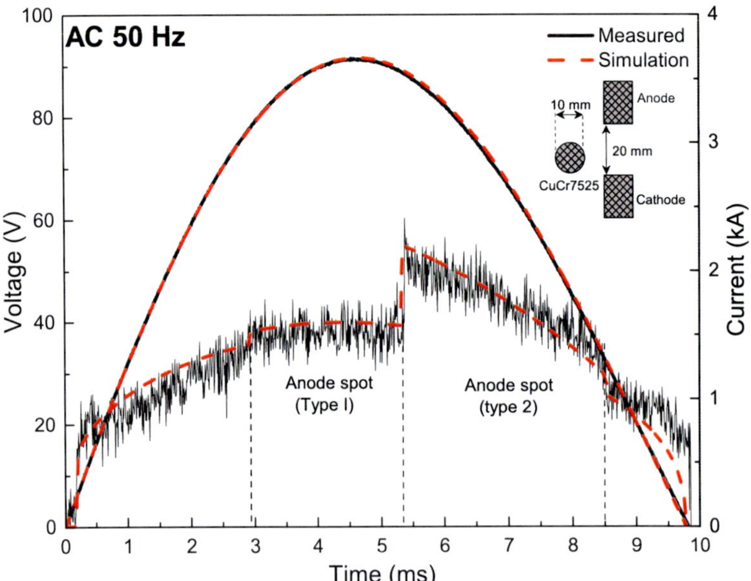


**Fig. 8.4** Measured and simulated current and voltage based on the new arc model and existence diagram for case 1.

During anode spot type 1 which appears at lower currents and gap lengths, the arc voltage increases from 35 V to 55 V which is tracked properly by the new model. By increasing the current and gap length at 3.5 kA and 6.5 mm, respectively, the anode spot type 2 appears. The voltage is increased about 50% and then it starts to decrease. This voltage is also traced by the model properly. Furthermore, the slope of decreasing the voltage during second anode spot is also traced satisfactorily by the new model, which confirms the current dependency of the voltage during this high-current mode.

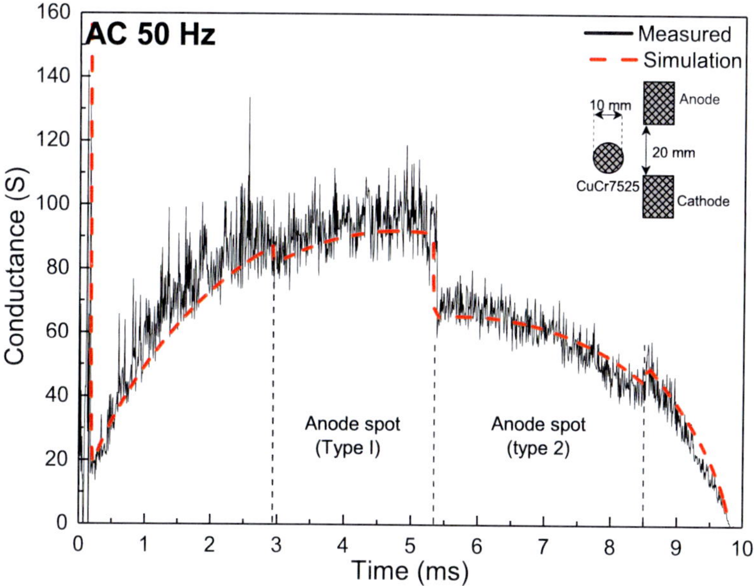


**Fig. 8.5** Measured and simulated electrical conductance based on new arc model and existence diagram for case 1.

The measured and simulated current and voltage related to the case 2 are presented in Fig. 8.6. In this case, the arc voltage during anode spot type 2 is traced perfectly by the model. During transition from footpoint to the first anode spot mode at 2.8 ms, the conductance is reduced by about 15% which is traced by the new model. By transition to the anode spot type 2 voltage rises by about 45% and subsequently it starts to decrease. This part is also predicted by the new model perfectly.

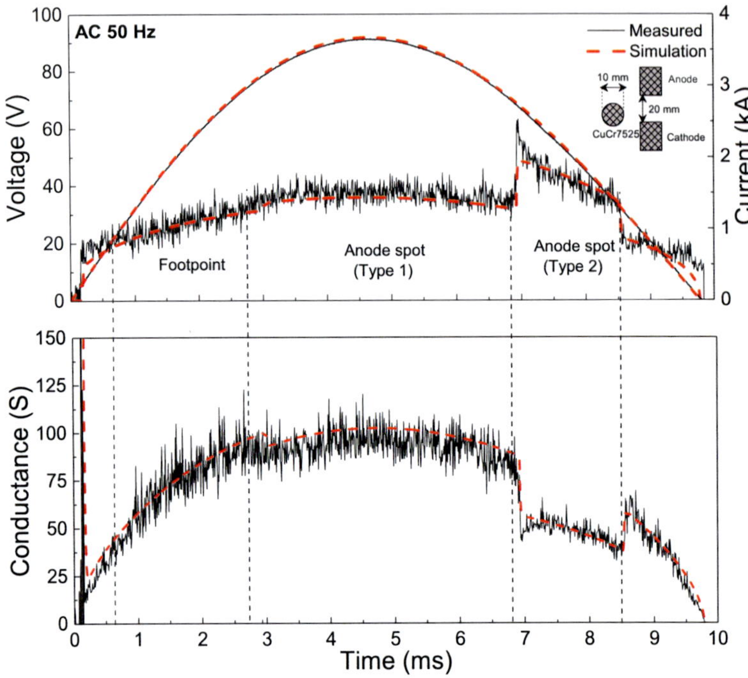


**Fig. 8.6** Measured and simulated current, voltage, and electrical conductance based on the new arc model for case 2.

## 8.4 Validation of the new model

The conventional arc models fail to trace the voltage or the conductance of the vacuum arc during high-current anode modes. The new arc model is developed based on the existence diagram which depends on waveforms, contact materials, and contact diameters, etc. Therefore, the existence diagrams and the boundaries between different modes have to be determined by performing sufficient amount of the shots. Although this causes a certain effort it should be taken into account that the calibration factor of the conventional arc models must be determined for different currents and setups as well [53–55].

In the conventional arc models, the arc parameters such as the arc voltage constant, the arc time constant as well as calibration factors have to be modified for each interruption current, electrode dimension and contact material. However, in the new arc model the arc voltage $U_h(t)$ is determined and it is valid for a quite wide range of interrupting currents. In fact, the new model based on the existence diagrams has superior prediction capability compared to the conventional arc model and calibration of all parameters are not necessary for different interrupting conditions.

In the following the validity of the new model is subjected to the investigation for AC 100 Hz and pulsed DC 10 ms.

## AC 100 Hz

The existence areas of high-current anode modes as a function of gap length and threshold current for AC 100 Hz and contact material of CuCr7525 with diameter of 10 mm are shown in Fig. 8.7. Two different displacement curves as examples corresponding to anode spot type 1 and 2 are presented by dotted and dashed lines, respectively. The approximate boundaries between different modes are illustrated by thick black lines.

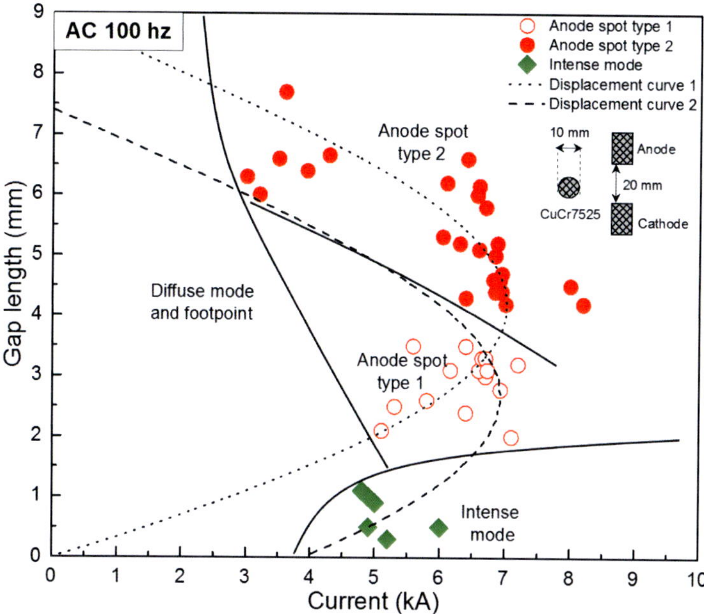


**Fig. 8.7** The existence areas of high-current anode modes as function of gap length and threshold current for AC 100 Hz and contact material of CuCr7525 with diameter of 10 mm.

The measured and simulated current, voltage, and electrical conductance for a vacuum arc with different discharge modes based on equation (8.1) and existence diagram for AC 100 Hz in Fig. 8.8. By transition to anode spot type 1 at 1.6 ms, the arc voltage increases slightly which is traced by the new model.

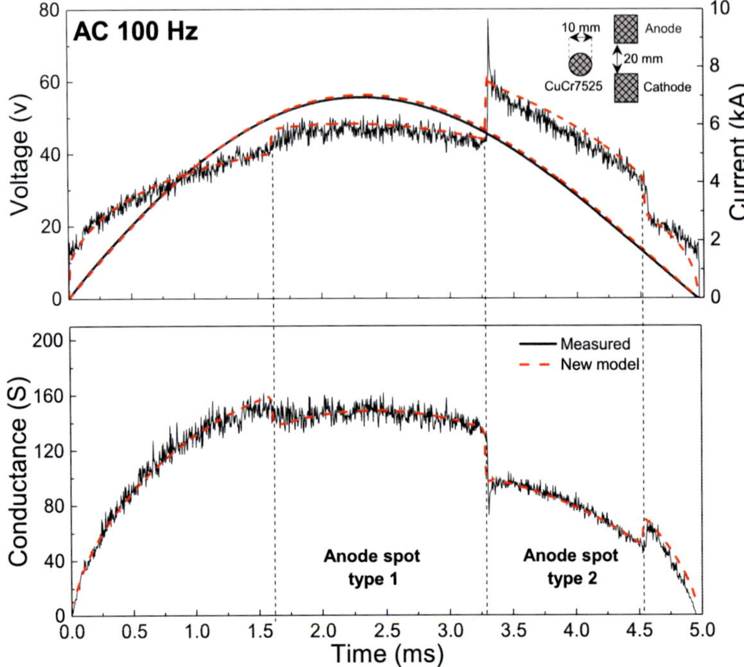


**Fig. 8.8** Measured and simulated current, voltage and electrical conductance based on the new arc model for AC 100 Hz.

By transition to anode spot type 2, the abrupt change in arc voltage appears at 3.3 ms which is in a good agreement with the presented results in [73, 92] and is traced properly by the new arc model (dashed line in Fig. 8.8).

## Pulsed DC 10 ms

The validity of the model is investigated by pulsed DC over 10 ms as well. The existence diagram of high-current anode modes for pulsed DC 10 ms and contact material of CuCr7525 with diameter of 10 mm is shown in Fig. 8.9.

The measured and simulated current and voltage for a vacuum arc are presented in Fig. 8.10 based on (8.1) and (2.4) for pulsed DC 10 ms. The results show that the voltage simulated by the new model (dashed lines) is rather similar to the measured one.

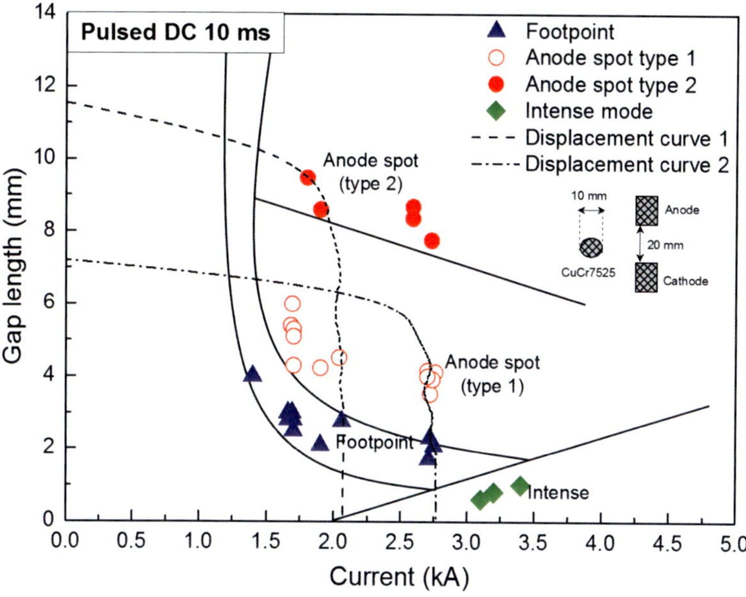


**Fig. 8.9** The existence areas of high-current anode modes as function of gap length and threshold current for pulsed DC 10 ms and contact material of CuCr7525 with diameter of 10 mm.

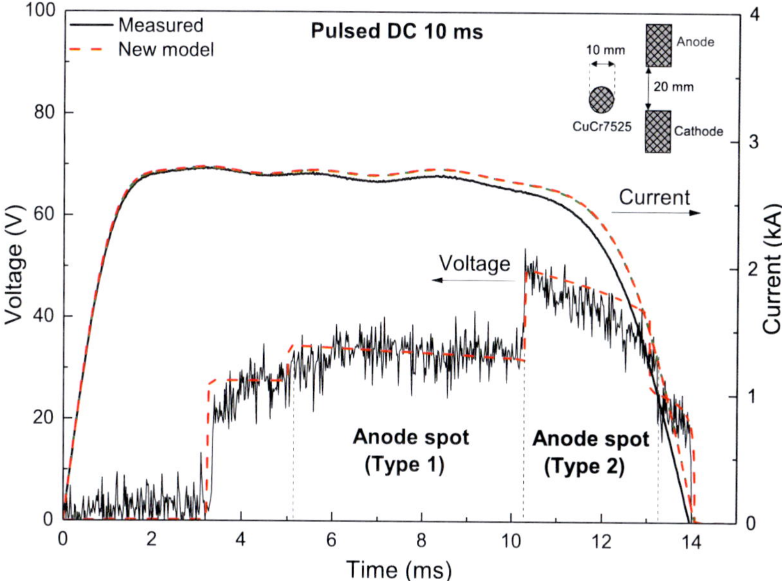


**Fig. 8.10** Measured and simulated current and voltage based on new arc model and conventional model for pulsed DC 10 ms.

## 8.5 Predictability of the model

Fig. 8.11 shows for another example the measured and simulated current, voltage, and electrical conductance during transition from diffuse mode to high-current modes based on the new model based on the parameter estimated for case 2 in Table 8.1. It seems that the voltage is traced by the model properly. However, the change in simulated arc voltage during transition to the anode spot type 1 starts earlier compared to the measured one. This can be explained by the threshold current of the anode spot type 1 which is presented in Fig. 8.12.

The threshold currents for footpoint, anode spot type 1 and type 2 modes for these two cases are presented again in Fig. 8.12. As it can be seen, the threshold currents of footpoint mode and type 1 in case 1 are near the border of the existence diagram, whereas the threshold current of anode spot type 1 in case 2 is a little away from border which leads to a little difference between simulation and measurement.

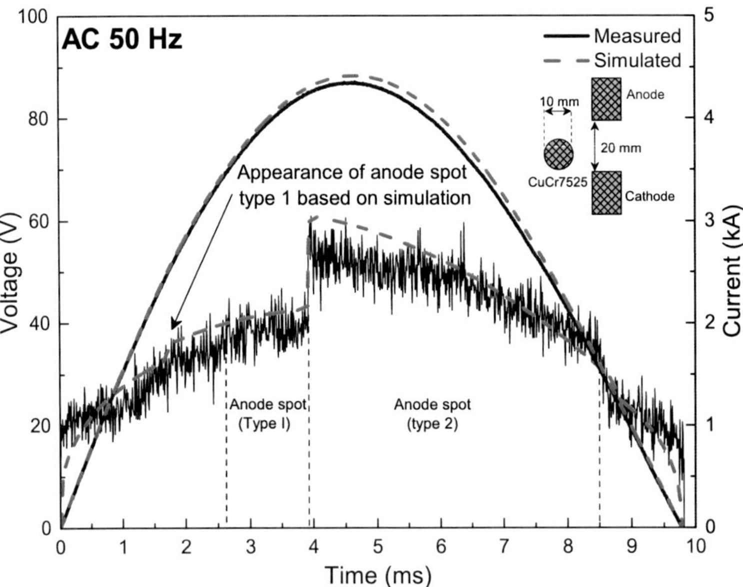


**Fig. 8.11** Measured and simulated current and voltage based on new arc model and existence diagram for case 2.

Although the distance between the threshold currents and boundaries in the existence diagram are changed, the results of the new arc model for vacuum arc based on the existence diagram is more precise in comparison with the conventional model.

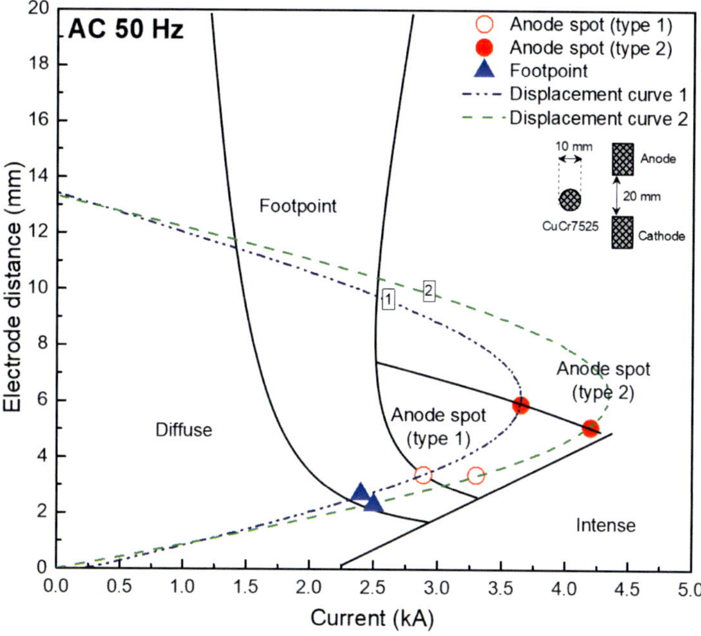


**Fig. 8.12** The threshold currents for footpoint, anode spot type 1 and type 2 modes for case 1 and 2 together with corresponding displacement curves.

Although the current is different in both cases presented in Fig.8,12, applying the same calibration parameters for both cases (cf. Table 8.1) the results are still acceptable. It confirms that the model has predictability properties. In contrast to conventional models in which the calibration factors must be modified for each test conditions, here no additional calibrations are required. In this regard the small difference between simulation and measurement is unavoidable (cf. Fig.8.11). Nevertheless, for more precise results the model parameters can be fitted for each experiments (cf. Fig. 8.4 to 8.6). It must be taken into account that the existence diagram depends on waveforms, contact materials, and contact diameters, etc.

# Chapter 9

# Discussion

## 9.1 Vacuum interrupter properties

Based on experiments conducted with AC and DC waveforms, the existence diagrams of the high-current anode phenomena with respect to threshold currents and gap lengths are studied systematically. This is carried out for different contact materials, gap geometries, and contact opening speeds.

By earlier opening time or increasing the contact opening speed, higher gap length can be achieved during discharge process which can lead to formation of the anode spot type 2 that is more active compared to the conventional anode spot. The impact of opening time or contact opening speed on the formation of anode spot type 2 is also investigated by existence diagram [73, 92, 94]. Moreover, it is proven in accordance with former studies that enlargement of electrode diameter as well as increase of the amount of Cu in the contact material increase the threshold currents at which mode transition occurs [31, 32, 92]. This is in contrast to the case with CuCr alloy, which has a higher interrupting capacity than Cu. Nevertheless, the addition of Cr dramatically reduces the thermal conductivity of the CuCr composite [1]. Moreover, the presence of metal droplets in the contact gap even after current zero can reduce the interrupting capability. Therefore, even though Cu has a higher threshold current at which first a high-current anode mode appears, it has a lower interrupting capacity than the CuCr family [1, 32]. This lies in the fact that the metal droplets of Cu will leave the surface after current zero, which evidently has a significant impact on the dielectric recovery strength of vacuum interrupters [15].

## 9.2 Comparison of DC and AC pulses

The impact of the current waveform regarding frequency and pulse duration on the formation of high-current anode modes is presented in Fig. 9.1 using transferred charge and threshold current. For this comparison AC 50 Hz, 100 Hz, 180 Hz, and 260 Hz as well as DC pulses over 5 ms and 10 ms are chosen. The CuCr7525 electrodes with diameter of 10 mm are used. The opening time is almost constant in all cases; however, the time in which the first high-current anode modes occurs is different for cases presented in Fig. 9.1. It means that the gap length is different for these cases.

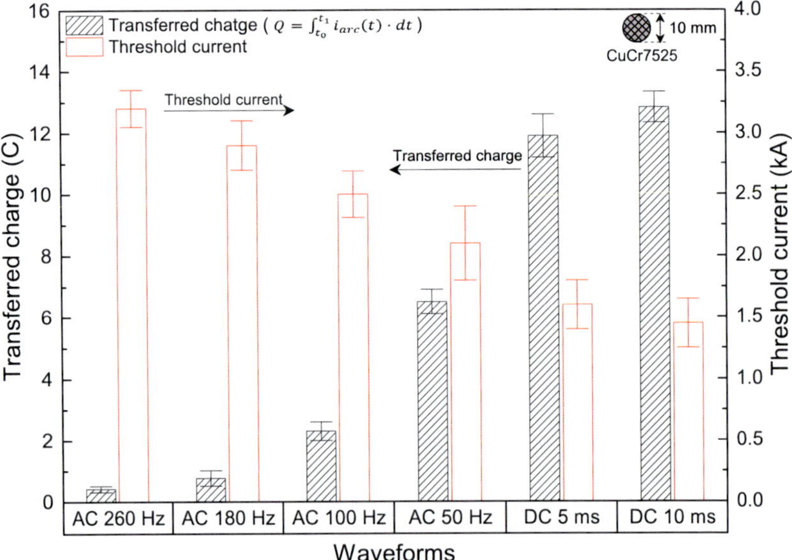


**Fig. 9.1** Transferred charge and threshold current of different waveforms for anode spot formation. The contact material CuCr7525 and the electrode diameter of 10 mm are selected. Contact speed is 2 m/s.

The results show that by increasing the frequency the threshold current of the formation of high-current anode modes increases, whereas by DC pulses the transferred charge increases.

When a high-current mode, e.g. anode spot forms within a few milliseconds in case of a high frequency current, more power is required than in the case of DC pulses in which the arcing time lasts for tens of milliseconds [9, 92]. The impact of pulse duration is also confirmed in chapter 4, Fig. 4.14 in which the threshold current of first observed high current mode for DC 10 ms is lower than that for DC 5 ms.

These inverse relationships between the input power to the anode and the average time delay until high-current anode mode formation underlines the importance of both instantaneous current and total transferred charge in formation of high-current anode modes [92, 103]. This can also explain the presented results in Fig. 4.8 in which the threshold currents of the high-current anode mode formation increase with ascending frequency.

The transferred charge is quite low for high frequency pulses. Therefore, the formation of high-current anode phenomena depends primarily on the critical current in these cases. At longer pulses, the threshold current of high-current anode phenomena correlates to some extent with the transferred charge, i.e. a higher transferred charge leads to a decrease of the threshold current.

Comparison of existence diagram of AC 50 Hz and DC 10 ms presented in Fig. 4.5 and Fig. 8.11 confirms that the threshold currents of anode spot type 1 and anode spot type 2 in case of the AC half-wave are higher than those in case of the pulsed DC. It is observed that the input power is lower, the higher the average time delay till high-current anode mode formation is. It emphasizes the importance of both instantaneous current and total accumulated energy (transferred charge) in the formation of high-current anode modes [91, 92].

The dependency of high-current anode phenomena formation on the threshold current and transferred charge of AC and DC pulses can also be examined with respect to the Cu line intensities near the anode. The relative intensities of atomic and ionic copper lines in the vicinity of the anode as a function of current for AC 50 Hz and DC 10 ms pulses presented in Fig. 9.2. For this purpose, three lines of Cu I 510.55 nm, Cu II 508.90 nm, and Cu III 516.89 nm are selected. The number of lines are increased by discharge time.

In case of AC 50 Hz, it becomes obvious that the considerable intensity increase of all lines during transition to the anode spot type 2 happens after current maximum. The later decrease of the ionic line intensities with falling current and the fast decrease of the Cu III line can be explained by the required power input for maintaining the ionization level. Therefore, the relative intensity of the atomic line remains high even at decreasing current course.

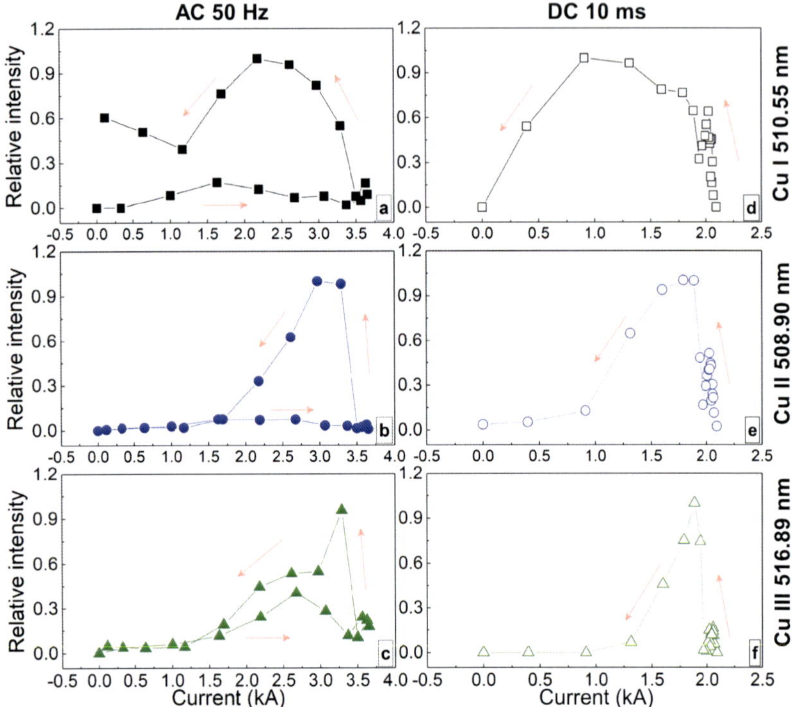


**Fig. 9.2** The relative intensity of atomic and ionic copper lines near the anode with respect to AC 50 Hz current for Cu I 510.55 nm (a), Cu II 508.90 nm (b), and Cu III 516.89 nm (c).

In case of the DC 10 ms, although the maximum current of the pulsed DC is higher than 2 kA, the anode spot type 2 occurs at lower currents, which confirms the impact of transferred charge on the formation of high-current anode phenomena in case of long pulses. In case of AC 50 Hz half-wave, after the current reaches maximum value, the anode spot type 2 appears in less than 1 ms; however, in case of the pulsed DC, the anode spot type 2 occurs after certain time delay, which can be discerned from the comparison of Fig. 9.2a and d. For high frequency pulses, the formation of high-current anode phenomena depends majorly on a critical current and is slightly affected by the total energy. At longer pulses, high-current anode phenomena formation depends on the transferred charge which is in good agreement with magnetic effects for short pulses and thermal effects for long pulses presented in [9].

## 9.3 Transition between different discharge modes

During the formation of the second mode, an abrupt change in arc voltage is recorded. Moreover, the high radiation intensity in the vicinity of both electrodes observed in the second type of anode spot mode corresponds to the characteristics of the intense mode [94]. However, the intense mode appears typically at lower gap lengths. Based on the existence diagrams of different high-current anode modes presented in [8, 12, 92], the second type of anode spot mode does not fit for the intense mode in the existence area. In addition, the arc voltage during intense mode is reported to be lower than the voltage during anode spot mode. Therefore, the mode observed at larger gap lengths is defined here as "anode spot type 2" in which both anode and cathode are active [92, 94].

The distribution of atomic and ionic copper lines near the anode, cathode and inter-electrode gap during and between transition of different discharge anode modes, i.e. diffuse, footpoint, anode spot type 1 and 2 modes are examined. The results show that during transition from footpoint to anode spot type 1 an abrupt change in the distribution of Cu III lines occur near the anode, the cathode and in the inter-electrode gap [93, 94].

The intensity of Cu III lines increases near both anode and cathode during anode spot; however, the intensity of Cu II lines is extremely decreased between the electrodes. This phenomenon can be related to the dark area in the inter-electrode gap during the formation of anode spot. During transition from anode spot type 1 to anode spot type 2, the relative intensities of atomic and ionic lines increase at both anode and cathode, while the intensity of atoms and ions is extremely low within the inter-electrode gap.

The most noticeable change is found for the line emission of the double charged ion radiation. In fact, higher ionization states e.g. Cu III can better define the arc dynamics during transition of high-current anode mode than atomic or the ionic line with lower ionization level. A pronounced dynamic behavior is found for the atomic lines as well, i.e. it must be expected that atoms play an active role in the high-current anode modes together with the ions.

Significant deviations from thermal and chemical equilibrium can be expected in the vacuum arc and in particular in the region of anode spots. Line intensities are related to the local densities of excited atoms and ions as well as to the local electron temperature. Exploring both the electron temperature and the total densities of atoms and ions with different charge numbers require a comparison with a detailed non-equilibrium plasma model. However, stronger radiation intensity of Cu III lines can be used as an indicator for regions of higher electron temperature and higher power density in the plasma needed for two-fold ionization of copper [73, 94].

## 9.4 Transition to anode spot type 2

A second type of anode spot modes as anode spot type 2 differs in the voltage course and the radiation intensity of atomic and ionic lines compared to other anode modes [73]. It shows certain similarities with the intense mode but occurs usually at larger electrode distances and sometimes even at lower threshold currents compared to the anode spot type 1. An abrupt change in the arc voltage appears during the transition from anode spot type 1 to anode spot type 2. In what follows, the formation of anode spot type 2 after anode spot type 1 is examined regarding transferred charge, aspect ratio, temporal evolution of atomic and ionic copper lines.

### Transferred charge

It is expected that the formation of high-current anode modes is governed by the transferred charge for longer pulses [73, 92, 103]. The transferred electric charge [103] is the time integral of the current during the arcing time. Fig. 9.3 represents the transferred charge and electrode travel curves for two cases presented in Fig. 4.9 with current of 2.7 kA and almost the same opening time and with different contact speeds. In case 2 at about 10 ms, the anode spot type 2 appears, where the transferred charge and gap length are about 19 C and 8.6 mm, respectively.

Considering the different opening times (300 $\mu$s), the same transferred charge (19 C) for case 1 can be reached at time instant and gap length of 10.4 ms and 5.9 mm. However, the anode spot type 2 does not appear in case 1. This comparison confirms that the formation of anode spot type 2 is neither explicitly dependent on the transferred charge nor on the instantaneous current. It can be concluded that the heat input to the surface is not the initiating moment for the anode spot mode formation [73]. It is noteworthy to mention that the current can affect the gap length in which anode spot type 2 appears (see Fig. 4.15b).

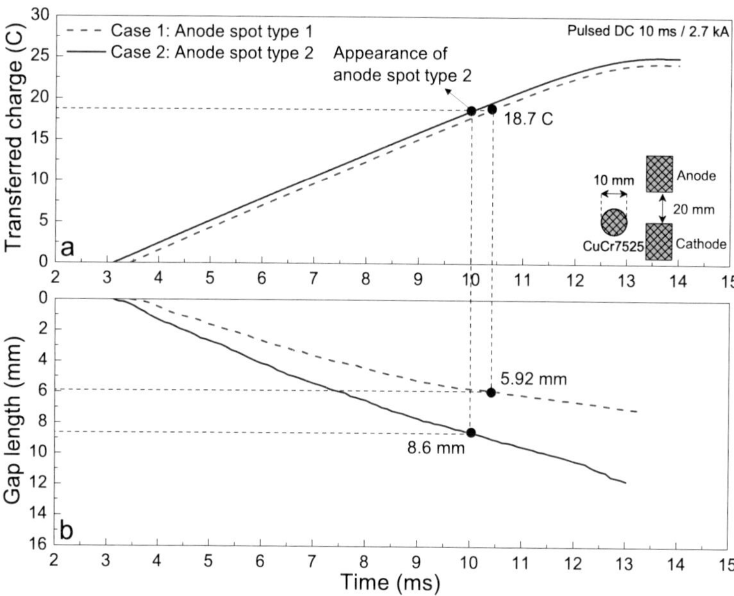


**Fig. 9.3** The transferred charge and electrode travel curves for two cases with the same current and almost the same opening time but with different opening speed [73].

## Aspect ratio

As discussed earlier, a certain amount of transferred charge and instantaneous current are not sufficient criteria for the formation of anode spot type 2. The aspect ratio which is the ratio of electrode diameter $D$ to gap length $l$ [73, 104, 105] should be considered as an additional parameter to investigate the formation of high-current anode modes. Fig. 9.4a and Fig. 9.4b present the aspect ratio as a function of the threshold current and the transferred charge from opening up to the formation of anode spot modes, respectively. Two currents peak of 2.1 kA and 2.7 kA are considered.

At both currents by decreasing the aspect ratio the anode spot type 2 appears. Nevertheless, by increasing the current from 2.1 kA to 2.7 kA, the aspect ratio in which anode spot type 1 or type 2 appear will be increased which is in a good agreement with results provided in [104, 105].

Fig. 9.4b confirms that the inset of the anode spot type 1 usually occurs with transferred charge of about 4 to 10 C (red unfilled circles), while the anode spot type 2 in some cases appears at a transferred charge higher than 13 C (red filled circles). However, the transferred charges can reach values above 20 C during the current pulse (red filled triangles) without transition to anode spot type 2.

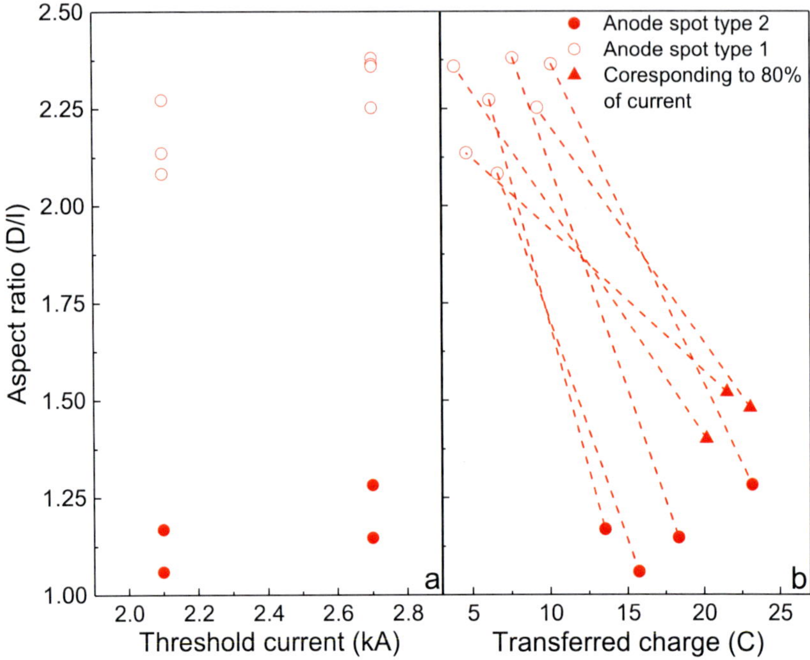


**Fig. 9.4** The aspect ratio (ratio of electrode diameter to gap length) as a function of transferred charge and threshold current of the according anode spot type 1 and type 2. [73].

To determine the transferred charge for cases without transition to anode spot type 2, the transferred charge is calculated up to 80% of the maximum current. The level of 80% is selected with respect to the appearance of anode spot type 2 which is depicted in existence diagram (Fig. 4.15). In fact, the corresponding time is the latest time at that anode spot type 2 can be appeared. Consequently, the transferred charge cannot serve as sufficient criterion to predict the formation of anode spot type 2. Moreover, it can be deduced that a certain value of aspect ratio is required for the formation of the anode spot type 2.

The relative intensities of atomic and ionic lines near the anode for case 1 and 2 as a function of the aspect ratio are presented in Fig. 9.5. The maximum current equals 2.1 kA. As it can be seen, at an aspect ratio of 1.16, the relative intensity of all species become bigger. Regarding Fig. 9.5, the maximum aspect ratio for transition from anode spot type 1 to anode spot type 2 is about 1.16 for current of 2.1 kA.

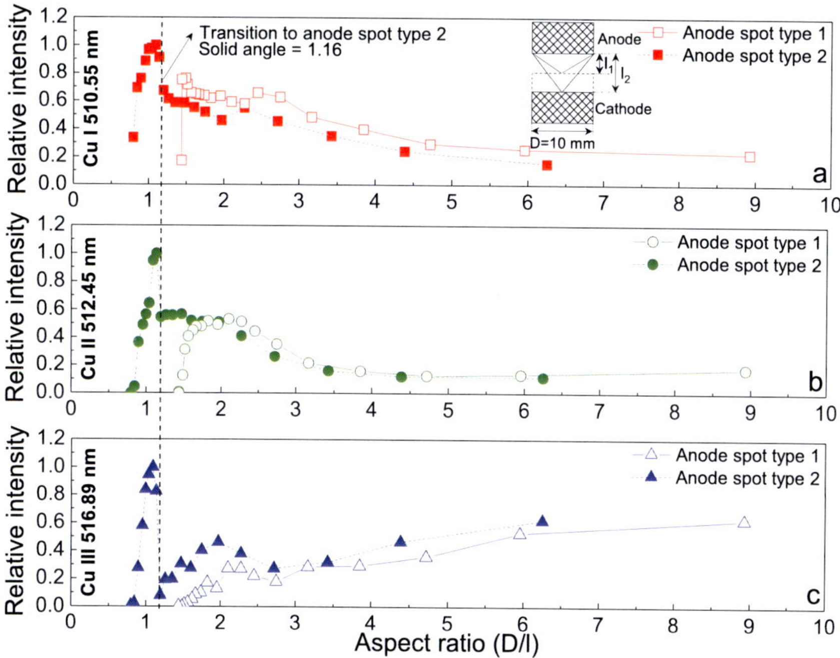


**Fig. 9.5** The relative intensity of Cu I, Cu II, and Cu III near the anode for both cases as a function of the aspect ratio. The maximum current reaches the value of 2.1kA [73].

## Summary

Different characteristics of the anode spot type 1 and type 2 can be summarized as follow:

- The spatial distribution patterns of Cu I, Cu II, and Cu III lines are similar in anode spot types 1 and 2.
- The intensities of Cu I, Cu II, and Cu III lines close to the anode in anode spot type 2 are much higher than those of anode spot type 1.
- An abrupt change in the spatial distribution of Cu III lines is observed at transition to anode spot types 1 and 2.
- A jump in the arc voltage appears at the transition from anode spot type 1 to type 2.
- The abrupt changes of optical emission and arc voltage emphasizes at a fast transition from one mode to another.

- The transition from anode spot type 1 to type 2 is governed mainly by the ratio of electrode diameter to gap length. However, increasing the current can reduce the value necessary for the formation of anode spot type 2.

A kind of mode transition during change from anode spot type 1 to type 2 can be interpreted roughly from an energetic point of view. The increased evaporation of electrode material due to local heating during the anode spot mode leads to a cloud of high density vapor in front of the anode. Such a cloud introduces a barrier for collision partners.

The higher required energy for excitation and ionization of the metal vapor together with the high radiation losses increases in essence the field strength required to drive the arc current. Hence, from an energetic point of view it might be preferable that the arc attachment is displaced towards colder regions of the anode surface with less evaporation in front of it. This would lead step by step to a continuous heating of the whole anode surface.

Finally, this gives rise to a qualitative change of the discharge mode when the whole anode surface is involved and the current must be transported through a region of relatively dense vapor. Short time instants of reduced intensity observed in the video frames presented in Fig. 9.6 (see dark regions in Fig. 9.6e and g) are typically accompanied by a voltage increment immediately before transition from anode spot type 1 to type 2, which might support the abovementioned explanation. In such short phases, elevated voltage and energy are required to ionize denser metal vapor regions.

The formation of anode spot can be explained by one or combination of three different models described in [9], i.e. changes in the inter-electrode plasma, changes in the anode surface and interaction between the anode surface and the adjacent plasma.

However, according to the results presented in chapter 5 and 7, the evaporation of anode surface appears during the formation of anode spot type 1 or type 2. The anode spot formation can be triggered by the evaporation of the anode material [30, 73]. In this model the preheating of the anode surface considered as a mechanism of the formation of anode spot. Moreover, the constriction occurs at the cold parts of the anode surface since the effect of the self-magnetic field can result in anode spot formation. In fact, the anode spot appears after the contraction.

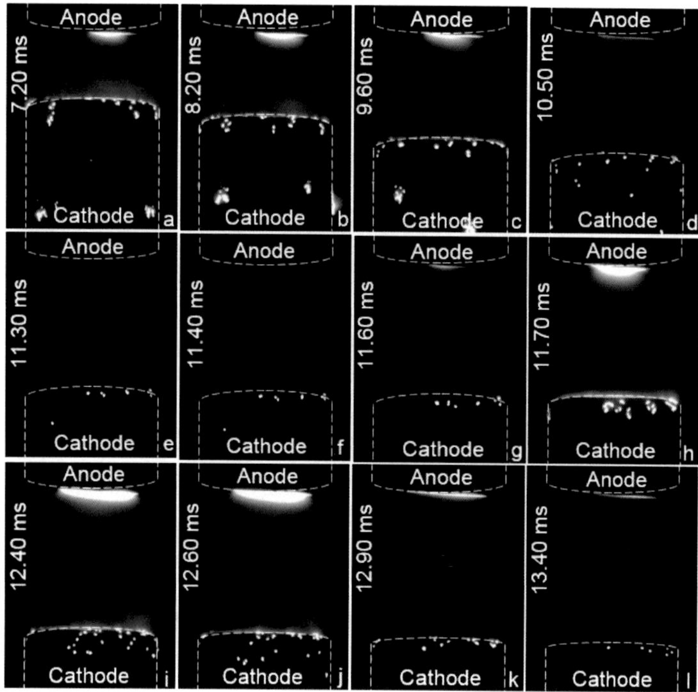


**Fig. 9.6** The transition from anode spot type 1 to type 2 in one shot [73].

## 9.5 Anode plume

A plasma plume is observed after extinction of anode spot type 2 and about 400 to 500 $\mu$s before current zero [41–43]. The plasma plume could be distinguished as an expanding vapor, which cloud is attached to the anode and surrounded by a bright shell [43]. The anode plume first contracts and later expands.

The distributions of copper atomic and ionic line intensities vertical and perpendicular to the electrodes confirm that the inner part of the anode plume is occupied with atoms and single-charged ions. However, the intensities of Cu II lines are low compared to Cu I lines inside the plume. The intensities of Cu II lines at the border of the anode plume are considerably higher than those of Cu I lines (see Fig. 6.10). The absence of Cu III lines reveals that the highly ionized plasma jet from cathode cannot be the source of the anode plume material. The bright shell of the anode plume with its increased amount of singly charged ions, however, will be a result of ionization. The anode plume is observed in case of CuCr electrodes, whereas no stable anode plume is observed in case of Cu electrodes [43, 82]. Moreover, the shape of anode plume is irregular in case of Cu electrodes and it follows liquid-metal protrusions and observed only above heat-insulated protrusions or around flying droplets which indicates that the surface evaporation rate plays a key role in the formation of the anode plume [42, 43]. The CuCr electrodes during the discharge behaves as a heat-insulated one [43].

It can be explained also by the immiscibility of copper and chromium even in the liquid phase [43]. Separate components have boundaries which can introduce obstacles in the heat transfer process. As a result of this, the surface becomes overheated, and the evaporation rate rises [43].

The anode plume can be explained as the result of a shock front formed in the discharge gap. It is caused by the interaction of the highly ionized plasma jet from the cathode with the atomic anode jet formed by evaporation that is expanding with thermal velocity [42, 43]. The plume shell is controlled by the balance between pressures in both the anode and cathode jets.

Different shapes for anode plume are introduced in chapter 6 (see Fig. 6.3, Fig. 6.4 and Fig. 6.9) which can be explained by different evaporation from the anode surface due to non-uniformity of surface temperature in a hot region of the anode [82, 43]. Moreover, the plume sizes are different shot by shot and as well as during one shot. This can be explained by the balance between the cathode plasma jet becoming less dense with the decrease in current and the anode jet whose density is kept high due to the anode heat capacity [42]. In fact, an anode plume shape contains information regarding the distribution of the evaporation rate of the anode surface.

The metal evaporation rate from anode surface is expected to play a key role in the plume appearance [42]. High metallic vapor density due to the anode plume before current zero can still exist even after current interruption; and it provides a potential source for arc reignition in the presence of transient recovery voltage [15].

During the extinction of anode spot type 2 and after abrupt decrease of the arc voltage, i.e. before current zero, an anode plume is established at the anode as shown in Fig. 9.7a. Fig. 9.7b pictures the phase at the time instant just before the formation of anode plume. Examination of many shots confirms that anode plume appears always and only after extinction of anode spot type 2 and before current zero. No anode plume can be observed in cases with only anode spot type 1. Fig. 9.7c shows the case with only anode spot type 1 and at the same time instant for the formation of anode plume. Anode plume is known as a source of vapor before current zero [73, 82] that is consistent with the atomic line in case 2 presented in Fig. 5.5 which is still high near the current zero depicted in Fig. 5.5b.

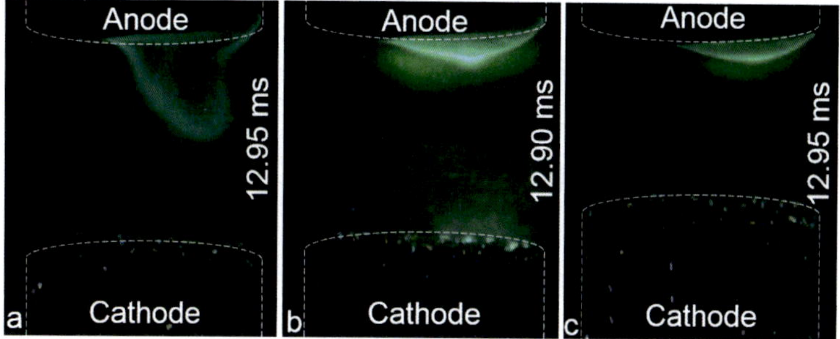


**Fig. 9.7** Formation of anode plume a) before current zero and after extinction of anode spot type 2, b) the time instant just before the formation of anode plume, and c) the same time instant of the formation of anode plume corresponding to the case in which only anode spot type 1 appears [73].

To investigate the optional of the reignition, Cr density in case with and without anode plume at 100 $\mu$s after current zero is presented as arithmetic mean of seven independent shots in Fig. 9.8. The Cr I transition at 425.43 nm is singled out for the comparison. Cr density in case without plume is significantly lower and about 60% of the Cr density in case with plume. The obtained Cr densities are about an order of magnitude lower than the Cu densities obtained by [76, 77] and orders of magnitude smaller than the critical value for Cu density of about $3\times10^{21}$ m$^{-3}$ given in [76]. However, a comparison of Cr densities to Cu densities is of less significance and is affected not only by the discharge conditions but also by the vapor pressures of both elements in the Cu-Cr alloy.

This work attempts for the first time at experimental investigation of Cr densities in high-current vacuum arcs. Therefore, no benchmark could be provided from literature.

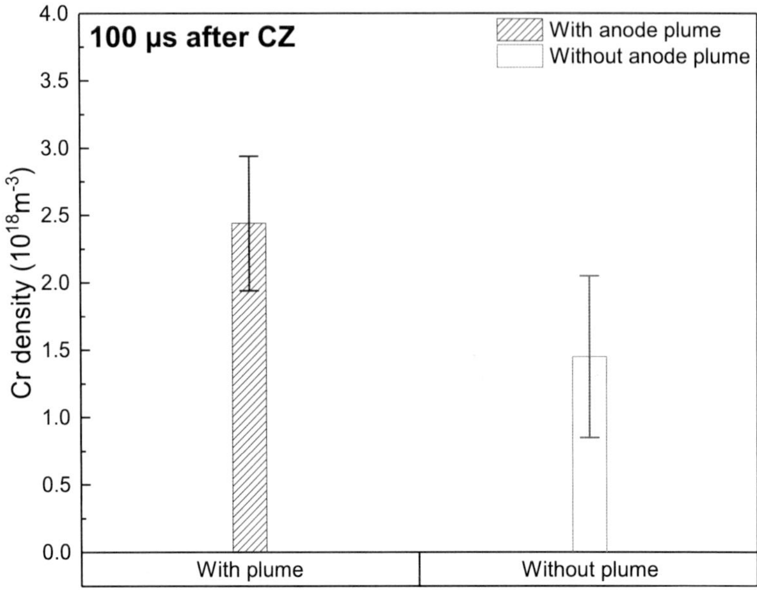


**Fig. 9.8** Cr density in case with and without anode plume 100 $\mu$s after zero crossing [76].

## 9.6 Specific electrical conductivity

According to the results presented in chapter 5, the radiation intensity in the case of the anode spot type 2 is higher in comparison with that of anode spot type 1. During the formation of the anode spot type 2, strong anode jets are formed, which probably interact with the cathode jets. The significant increase of the arc voltage during the anode spot type 2 should be due to the decrease of electrical conductivity. This could be expected to be caused by a layer of dense copper vapor in front of the anode. Spectroscopic analysis presented in chapter 5 confirms the growth of metal vapor in front of the anode surface during the formation of the anode spot type 2. The electrical conductivity of a case with anode spot type 2 is presented in Fig. 9.9. Considering the arc voltage during diffuse mode and anode spot type 1 mode, the specific electrical conductivity during anode spot type 1 can be calculated by using the arc voltage, temporal arc current, and the length and cross-sectional area of the arc as follow

$$u_{as1}(t) = u_{ca} + \sigma_{co}(t) \cdot \frac{l_{as1}(t)}{A_{as1}(t)} \cdot i(t) \tag{9.1}$$

where $u_{ca}$ is the arc voltage during diffuse mode, $\sigma_{co}$ is specific conductivity of the arc column; $u_{as1}$ stands for the arc voltage during anode spot type 1, $l_{as1}$ and $A_{as1}$ stand for the length and the cross section of the arc during anode spot type 1. The arc voltage during diffuse mode is typically in the range of 20-30 V.

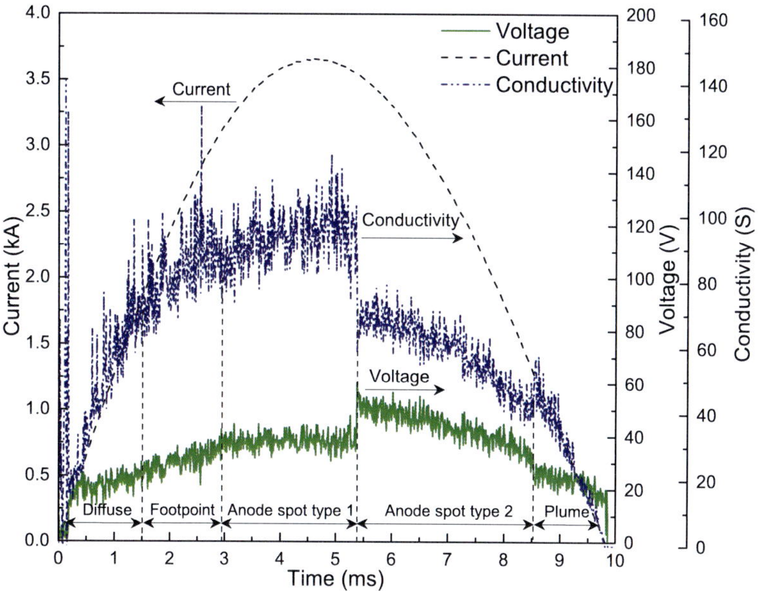


**Fig. 9.9** Electrical conductivity of a case with anode spot type 2 [92].

Moreover, the electrical conductivity during anode spot type 2 can be calculated by using the arc voltage during anode spot type 1 and 2 and geometric parameters of the arc as

$$u_{as2}(t) = u_{as1}(t) + \sigma_{as2}(t) \cdot \frac{l_{as2}(t)}{A_{as2}(t)} \cdot i(t) \tag{9.2}$$

where $u_{as2}$ is the arc voltage during anode spot type 2, $\sigma_{as2}$ is the specific conductivity during anode spot type 2, $l_{as2}$ and $A_{as2}$ are the length and the cross section of the are near the anode during spot type 2.

The specific electric conductivity based on the electrical representation in (9.1) and (9.2) during anode spot type 1 and 2 are indicated with unfilled and filled circles in Fig. 9.10, respectively.

Under the assumption that the vapor density has a dominant impact on the electrical conductivity, the specific electrical conductivity should depend on vapor density $N_a$, electron density $n_e$ and plasma temperature $T$ according to

$$\sigma \sim \frac{1}{\sqrt{T}} \frac{1}{\frac{N_a}{n_e}}. \tag{9.3}$$

Using the results for $N_a$ and $n_e$ presented in Fig. 5.8 and Fig. 5.9 a relative specific electrical conductivity has been estimated according to (9.3) which is illustrated in Fig. 9.10. Such an estimation coincides well with the change of the measured conductivity from anode spot type 1 to anode spot type 2 and yields also to some extent the decreasing conductivity during anode spot type 2.

Unfortunately, it must be stated that under the high ionization degree of the plasma (compare vapor and electron density in Figs. 5.8 and 5.9) a dominant effect of vapor atoms on the specific electrical conductivity can hardly be assumed, and the conductivity may be mostly equal to the Spitzer conductivity ($\sim T^{-3/2}$). However, in this case the measured evolution of the conductivity can not be explained in a simple way and remains an open question.

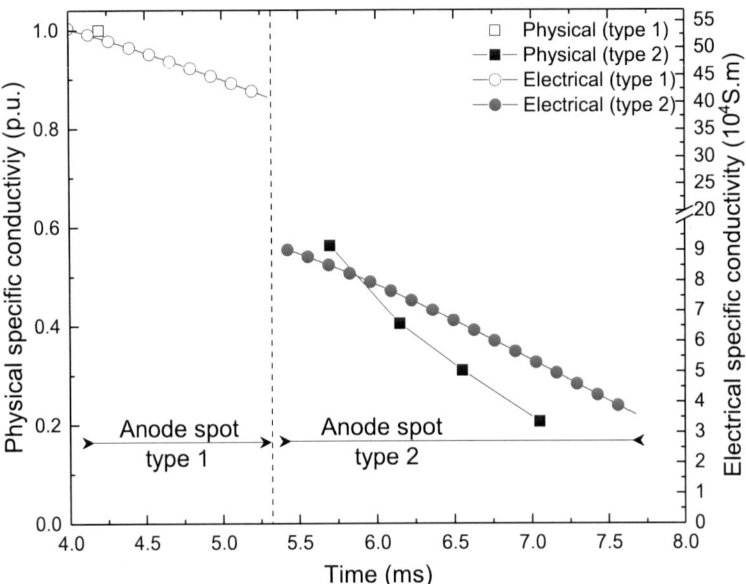


**Fig. 9.10** The specific electric conductivity according to the electrical and physical representation during anode spot type 1 and 2.

# Chapter 10

# Summary and outlook

The impact of several vacuum interrupter properties including electrode diameter, gap geometry, contact opening time and speed on the formation of high-current anode phenomena has been investigated systematically. Threshold currents together with corresponding gap lengths of the formation of high-current anode modes are determined for AC frequencies of 50, 100, 180, and 260 Hz and pulsed DC over 5 and 10 ms. The influence of various abovementioned parameters are investigated using existence diagrams. The results show that both instantaneous current and transferred charge are important for the formation of high-current anode phenomena. Nevertheless, the formation of high-current anode modes is dominated by current in case of high frequency pulses and by arcing time in pulsed DC.

A new type of high-current anode modes, anode spot type 2 is found and described for the first time. Anode spot type 2 can be distinguished from the known anode spot (type 1) and from the intense mode by its characteristic behavior concerning arc voltage and light emission near the electrodes. In contrast to anode spot type 1, both anode and cathode are active in case of anode spot type 2. Comparing to anode spot type 1 and intense modes, anode spot type 2 occurs at higher gap lengths. Moreover, an abrupt increase of 10-20 V in the arc voltage appears during transition to the anode spot type 2.

Video spectroscopy is used to investigate the distribution of atomic and ionic lines near the anode, the cathode, and in the inter-electrode. The results show a similarity in both AC and DC types of waveforms. Some unique changes appear during transition between different modes in both waveforms, e.g. an abrupt change in Cu III emission lines, is recorded during transition from footpoint mode to anode spot type 1 mode. Moreover, noticeable increases of the intensity of all species are observed during transition to anode spot type 2. Physical parameters of the plasma, e.g. radiating density, ground state density and electron density are determined during anode spot type 1 and type 2 using optical emission spectroscopy.

Transition to anode spot type 2 is investigated with respect to transferred charge, aspect ratio, existence diagram and distribution of atomic and ionic lines. In fact, a transition from anode spot type 1 to type 2 is dominated by the ratio of electrode diameter to gap length (aspect ratio). Nevertheless, increasing the current can decrease the aspect ratio which is necessary for the formation of anode spot type 2.

A unique consequence of anode spot type 2 is the possible appearance of another new phenomenon, the anode plume. It appears immediately after extinction of anode spot type 2, i.e. just the before current zero, and is characterized by a voltage decrease of about 10 V. Optical emission spectroscopy confirms that the anode plume consists of a cloud in the front of the anode surface formed by metal vapor from the anode that is covered by a shell of single charged ions. The radiating densities of Cu I and Cu II and the temporal evolution of the plasma temperature are determined inside the anode plume. Both density and temperature decrease by extinction of anode plume, i.e. under strong decrease of current. The formation of an anode plume is stable from shot to shot in case of CuCr electrodes but not in case of Cu electrodes.

Ground state densities of Cr I resonance lines are determined after current zero as well as in the active phase using broadband absorption spectroscopy. In case of extinction of anode spot type 2 the density around current zero of about $2.4\times10^{18}\,\mathrm{m}^{-3}$ decreases by about factor of five within the next 3 ms. During active phase, the density for anode spot type 2 is about 6 times higher than for anode spot type 1. The Cr ground state density during anode plume at about 200 $\mu$s before current zero is noticeably higher than at current zero. Moreover, the ground state density of Cr at about 100 $\mu$s after current zero is compared for two cases with and without anode plume. It is about two times higher in case of anode plume.

It is shown that conventional arc models, e.g. Mayr, Cassie, Schwarz, Schavemaker, and KEMA are not able to trace the arc voltage of high–current vacuum arcs, especially its abrupt change during transition between anode modes. Therefore, a new electric arc model is proposed which can trace the arc voltage of vacuum arc even during anode spot type 2. Besides, this model has predictability characteristics due to application of predictable existence diagram to the model. Thus, the arc voltage can be traced within quite different interrupting current ranges.

These results can be used to propose a physical model which can explain the formation of different high-current anode modes especially the new ones, i.e. anode spot type 2 and anode plume. It is expected that LTE conditions are not valid in case of high-current vacuum arc. So, a non-LTE model can be developed to explain high-current anode phenomena in more details from physical point of view.

In this work, to reduce the complicity of magnetic field, butt electrodes made out of CuCr with relatively low electrode diameter are used without external field. However, the final goal is to simulate the current density of real vacuum interrupters in which TMF or AMF contacts are widely used. So, using the TMF or AMF electrodes with higher electrode diameter can be considered as a future work.

# A. Experimental setup

Here, the experimental setup including high-current generator, vacuum chamber, and spectrograph are presented. Note, that additional information can be found in chapter 3.

**Fig. A1** Photograph of the high-current generator together with the vacuum chamber and the high-speed camera: 1) coils for pulsed DC, 2) coils for damped AC, 3) spark gap, 4) capacitor, 5) vacuum chamber, 6) high-speed camera, and 7) actuator.

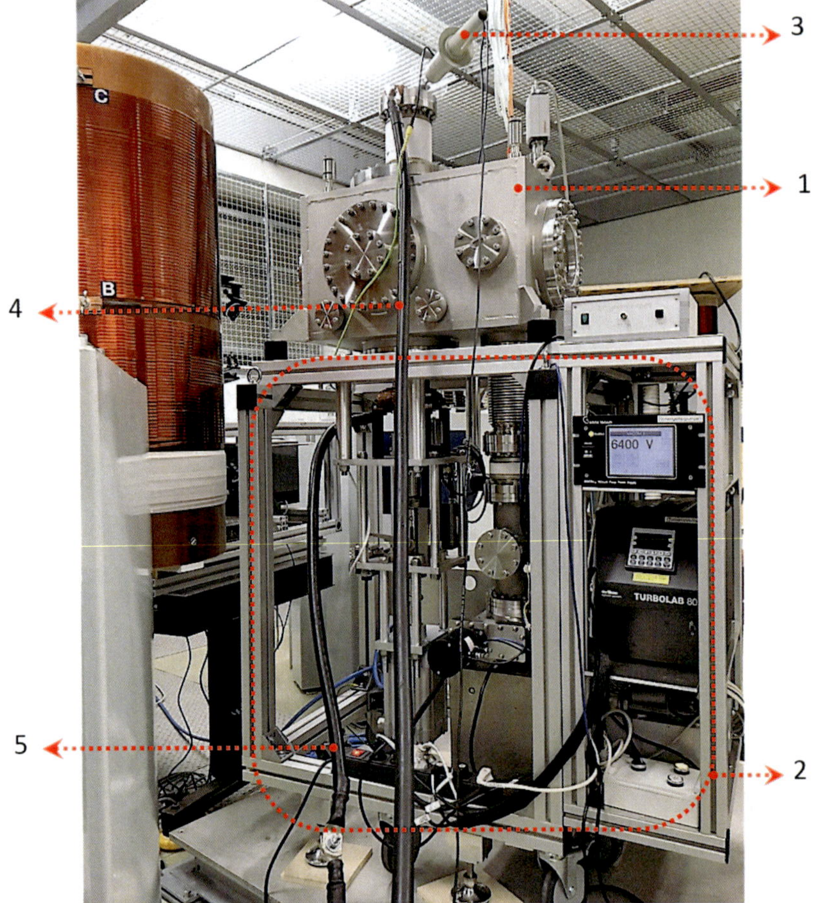


**Fig. A2** Photograph of the UHV chamber and pneumatic actuator: 1) Vacuum chamber, 2) actuator, 3) voltage probe, 4) HV cable, 5) ground.

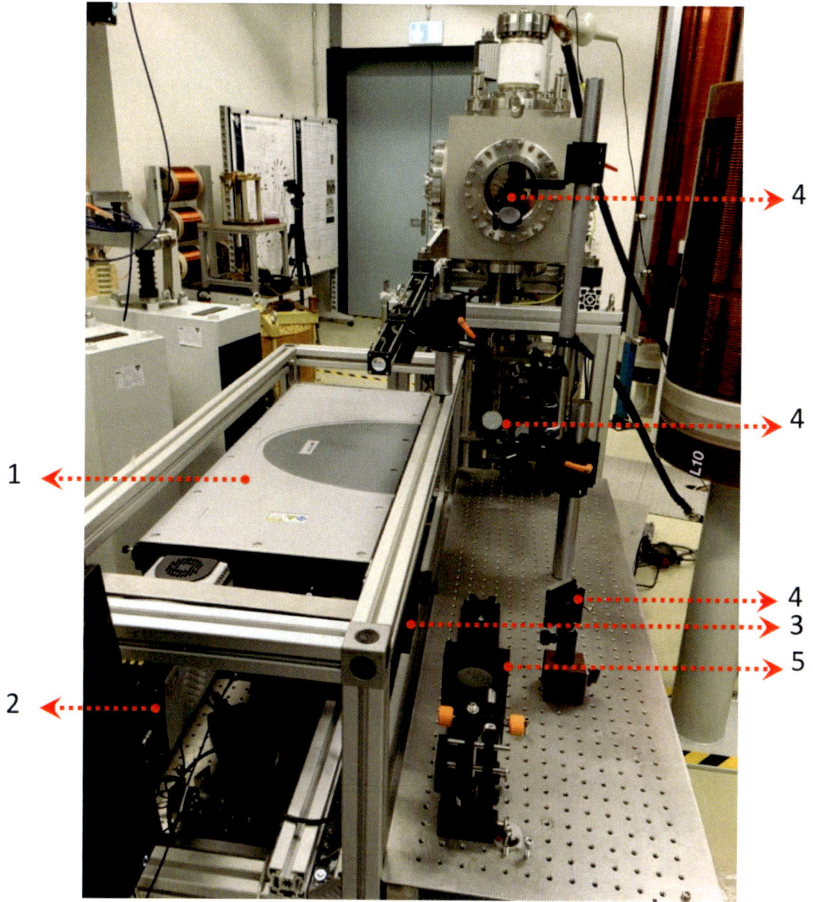


**Fig. A3** Photograph of the spectroscopy setup: 1) spectrograph, 2) ICCD camera, 3) entrance slit, 4) reflecting and 5) focussing mirrors.

**Fig. A4** Butt electrodes applied in this work: 1) cathode, 2) anode.

# B. List of symbols

| Symbol | Description | Unit |
|---|---|---|
| $\alpha$ | Constant parameter for arc diameter | |
| $\beta$ | Constant parameter for arc diameter | |
| $\dot{m}_{ax}$ | Axial mass flow (power based model) | $\mathrm{kg/ms}$ |
| $\dot{m}_{rad}$ | Radial mass flow (power based model) | $\mathrm{kg/ms}$ |
| $\varepsilon(i)$ | Weight function of Voronin-Sawiki arc model | |
| $\gamma$ | Half-width at half-maximum (HWHM) of the Lorentzian profile | nm |
| $\lambda_0$ | Center wavelength of the considered line | nm |
| $\lambda_i$ | Selecting of different arc model in Kema model | |
| $\lambda_{lu}$ | Transition line wavelength | nm |
| $\mu$ | Permeability | $\mathrm{N/A^2}$ |
| $\omega$ | Angular frequency | Hz |
| $\rho_{arc}$ | Mass density in the arc (power based model) | $\mathrm{kgm^{-1}}$ |
| $\sigma$ | Standard deviation of the Gaussian profile | |
| $\sigma_{as2}$ | Specific conductivity during anode spot type 2 | $\Omega\mathrm{mm^{-1}}$ |
| $\sigma_B$ | Stefan-Boltzmann constant | $\mathrm{Wm^{-2}K^{-4}}$ |
| $\sigma_{co}$ | Specific conductivity of the arc column | $\Omega\mathrm{mm^{-1}}$ |
| $\tau_c$ | Arc time constant of Cassie arc model | $\mu\mathrm{s}$ |
| $\tau_d$ | Arc time constant of arc model based on the arc diameter | $\mu\mathrm{s}$ |
| $\tau_i$ | Arc time constant of n-th arc model | $\mu\mathrm{s}$ |
| $\tau_{mm}$ | Arc time constant of modified Mayr arc model | $\mu\mathrm{s}$ |

| | | |
|---|---|---|
| $\tau_m$ | Arc time constant of Mayr arc model | $\mu$s |
| $\tau_s$ | Arc time constant of Schavemaker arc model | $\mu$s |
| $\tau_{vs}$ | Arc time constant of Voronin-Sawiki arc model | $\mu$s |
| $\theta$ | Breakdown voltage constant | |
| $\varepsilon_{lu}$ | Local emission coefficient | $\mathrm{Wm^{-3}sr^{-1}}$ |
| $a$ | Dependencies of arc time constant from the arc conductance | |
| $A_{arc}$ | Arc cross section (power based model) | $\mathrm{mm^2}$ |
| $A_{as1}$ | Cross section of the arc during anode spot type 1 | $\mathrm{mm^2}$ |
| $A_{as2}$ | Cross section of the arc during anode spot type 2 | $\mathrm{mm^2}$ |
| $A_b$ | Breakdown voltage constant | |
| $A_c$ | Effective cross-section area of the core | $\mathrm{m^2}$ |
| $a_d$ | Constant for arc model based on the arc diameter | |
| $a_e$ | Calibration parameter of arc model based on the existence diagram | |
| $A_i$ | Cooling constant of n-th arc model | $\mathrm{W^{-1}}$ |
| $A_{lu}$ | Transition probability | $\mathrm{s^{-1}}$ |
| $B$ | Magnetic flux density | $\mathrm{Nm^{-1}A^{-1}}$ |
| $b$ | Dependencies of dissipated power from the arc conductance | |
| $B_\lambda$ | Calibration curve of tungsten strip lamp | $\mathrm{W/nmm^2sr}$ |
| $b_e$ | Calibration parameter of arc model based on the existence diagram | |
| $C$ | Capacitance | F |
| $c$ | Speed of light in vacuum | m/s |
| $C_0$ | Capacitance of discharge circuit of xenon lamp | mF |
| $C_c$ | Capacitance of measuring cable | mF |
| $c_d$ | Constant for arc model based on the arc diameter | |
| $c_e$ | Coefficient of contact opening speed | |
| $C_{HV}$ | High voltage capacitance of the voltage divider | mF |
| $C_{in}$ | Capacitance of each stage of INP's current generator | $\mu$F |
| $C_i$ | Current constant of modified Mayr arc model | V |
| $C_{LV}$ | Low voltage capacitance of the voltage divider | mF |

| | | |
|---|---|---|
| $D$ | Effective absorption length | mm |
| $d(i)$ | Arc diameter of Voronin-Sawiki arc model | mm |
| $d_1$ | Constant for formula 1 for arc diameter | mm |
| $d_2$ | Constant for formula 2 for arc diameter | mm |
| $d_b$ | Gap length at breakdown | mm |
| $d_e$ | Diameter of the xenon lamp | cm |
| $d_k$ | Arc diameter of arc model based on the arc diameter | mm |
| $e$ | Elementary charge | C |
| $E_u$ | Excitation energy of the upper state | eV |
| $F_{DC}$ | Transfer function of capacitive part of the voltage divider | |
| $F_{DR}$ | Transfer function of resistive part of the voltage divider | |
| $f_{lu}$ | Oscillator strength | |
| $G(x;\sigma)$ | Gauss function | |
| $g_c$ | Conductance of Cassie arc model | S |
| $g_d$ | Conductance of arc model based on the arc diameter | S |
| $g_e$ | Conductance of arc model based on the existence diagram | S |
| $g_i$ | Conductance of n-th arc model | S |
| $g_k$ | Conductance of Kema arc model | S |
| $g_{mm}$ | Conductance of modified Mayr arc model | S |
| $g_m$ | Conductance of Mayr arc model | S |
| $g_{sw}$ | Conductance of Schwarz arc model | S |
| $g_s$ | Conductance of Schavemaker arc model | S |
| $g_u$ | Statistical weight of the excited upper state | |
| $g_{vs}$ | Conductance of Voronin-Sawiki arc model | S |
| $H$ | Magnetic field | $Am^{-1}$ |
| $h$ | Planck constant | $m^{-2}kgs^{-1}$ |
| $h_{arc}$ | Enthalpy in arc (power based model) | $J/kg$ |
| $h_{av}$ | Temporal average enthalpy of arc (power based model) | $Jkg^{-1}s^{-1}$ |

| | | |
|---|---|---|
| $I$ | Transmitted spectral intensity | Counts |
| $i$ | Arc current | A |
| $I_0$ | Spectral intensity outside the absorption profile | Counts |
| $i_{max}$ | Maximum current | A |
| $i_p$ | Primary current of current transformer | A |
| $I_p(s)$ | Laplace transform of primary current of current transformer | A |
| $i_s$ | Secondary current of current transformer | A |
| $I_s(s)$ | Laplace transform of secondary current of current transformer | A |
| $k(\lambda)$ | Absorption coefficient | $mm^{-1}$ |
| $k_{axmf}$ | Calibration constant of power gain carried by axial mass flow | |
| $k_B$ | Boltzmann constant | $m^2kg/s^2K$ |
| $K_g$ | Coefficient of approximation of unitary conductance | |
| $k_{net}$ | Calibration constant of net power of the arc | |
| $k_{rad}$ | Calibration constant of radiative power loss | |
| $k_{ramf}$ | Calibration constant of power loss carried by radial mass flow | |
| $k_{tur}$ | Calibration constant of turbulent power loss | |
| $L$ | Inductance | H |
| $L_0$ | Inductance of discharge circuit of xenon lamp | mH |
| $L_{\lambda P}$ | Spectral radiance | $W/nmm^2sr$ |
| $L_\lambda$ | Background light source (measured without plasma) | Counts |
| $l_{arc}$ | Arc length (power based model) | mm |
| $l_{as1}$ | Length of the arc during anode spot type 1 | mm |
| $l_{as2}$ | Length of the arc during anode spot type 2 | mm |
| $l_c$ | Mean path length within the core of current transformer | m |
| $L_{DAC}$ | Inductance of each stage of INP's current generator | $\mu H$ |
| $l_e$ | Length of the xenon lamp | cm |
| $L_{in}$ | Inductance for various pulsed DC of INP's current generator | $\mu H$ |

| | | |
|---|---|---|
| $l_{vs}$ | Arc length of Voronin-Sawiki arc model | mm |
| $L_{ZUS}$ | Inductance for various frequencies of INP's current generator | $\mu$H |
| $m_e$ | Electron mass | kg |
| $n$ | Number of the coil's turn of current transformer | |
| $N_a$ | Vapor density | $\mathrm{m}^{-3}$ |
| $n_e$ | Electron density | $\mathrm{m}^{-3}$ |
| $n_u$ | Radiating density | $\mathrm{m}^{-3}$ |
| $P$ | Dissipated power at current zero of Schwarz model | W |
| $P_0$ | Power loss of arc | W |
| $P_1$ | Cooling constant of Schavemaker arc model | |
| $P_\lambda$ | Plasma radiation (measured without background) | Counts |
| $p_{arc}$ | Arc pressure (power based model) | bar |
| $P_{axmf}$ | Power gain carried by axial mass flow | W |
| $p_{mm}$ | Pressure of circuit breaker of modified Mayr arc model | bar |
| $P_{net}$ | Net power of the arc | W |
| $P_{rad}$ | Radiative power loss | W |
| $P_{ramf}$ | Power loss carried by radial mass flow | W |
| $P_S$ | Power dissipated through the lateral surface of the arc column | W |
| $P_{total}$ | Total power balance of the arc | W |
| $P_{tur}$ | Turbulent power loss | W |
| $P_V$ | Power dissipated through the volume of the arc column | W |
| $PL_\lambda$ | Radiation of the plasma and the background light source | Counts |
| $q$ | Electric load | C |
| $Q(T)$ | Partition function | |
| $q_{max}$ | Maximum electric load | C |
| $R$ | Resistance | Ω |
| $r_{arc}$ | Arc radius (power based model) | mm |
| $R_e$ | Resistance of the discharge in tube | Ω |
| $R_{HV}$ | High voltage resistance of the voltage divider | MΩ |

| | | |
|---|---|---|
| $R_{LV res}$ | Resulting low voltage resistance of the voltage divider | MΩ |
| $S_{\lambda P}$ | Measured signal of plasma | Counts |
| $S_{\lambda WBL}$ | Measured signal of tungsten strip lamp | Counts |
| $T$ | Plasma temperature | K |
| $T_{arc}$ | Arc temperature (power based model) | K |
| $T_{DC}$ | Duration of pulsed DC | s |
| $t_P$ | Exposure times of the measured signals of plasma | s |
| $t_{WBL}$ | Exposure times of the measured signals of tungsten strip lamp | s |
| $u$ | Arc voltage | V |
| $U_0$ | Arc voltage during diffuse mode | V |
| $u_0$ | Average arc voltage of Cassie arc model | V |
| $U_{arc}$ | Average arc voltage of Schvemaker arc model | V |
| $u_{as1}$ | Arc voltage during anode spot type 1 | V |
| $u_{as2}$ | Arc voltage during anode spot type 2 | V |
| $U_b$ | Breakdown voltage | V |
| $u_{ca}$ | Arc voltage during diffuse mode | V |
| $U_h(t)$ | Voltage coefficient related to different high-current modes | V |
| $V(x;\sigma,\gamma)$ | Voigt function | |
| $v_e$ | Contact speed | m/s |
| $v_{sound}$ | Speed of sound in the arc | m/s |
| $x$ | Shift from the line center | nm |
| $Z$ | Wave impedance | Ω |

# C. References


[1] P. G. Slade. *The Vacuum Interrupter Theory, Design, and Application.* CRC Press, Boca Raton, 2008.

[2] R. L. Boxman, D. M. Sanders, P. J. Martin, and J. M. Lafferty. *Handbook of Vacuum Arc Science and Technology Fundamentals and Applications.* NJ: Noyes, Park Ridge, 1995.

[3] R. L. Boxman, S. Goldsmith, and A. Greenwood. Twenty-five years of progress in vacuum arc research and utilization. *IEEE Trans. Plasma Sci.*, 25(6):1174–1186, 1997.

[4] E. Dullni and E. Schade. Investigation of high-current interruption of vacuum circuit breakers. *IEEE Trans. Dielectr. Electr. Insul.*, 28(4):607–620, 1993.

[5] E. Schade and E. Dullni. Recovery of breakdown strength of a vacuum interrupter after extinction of high currents. *IEEE Trans. Dielectr. Electr. Insul.*, 9(2):207–215, 2002.

[6] E. Schade. Physics of high-current interruption of vacuum circuit breakers. *IEEE Trans. Plasma Sci.*, 33(5):1564–1575, 2005.

[7] H. C. Miller. Vacuum arc anode phenomena. *IEEE Trans. Plasma Sci.*, 11(2):76–89, 1983.

[8] H. C. Miller. Discharge modes at the anode of a vacuum arc. *IEEE Trans. Plasma Sci.*, 11(3):122–127, 1983.

[9] H. C. Miller. A review of anode phenomena in vacuum arcs. *IEEE Trans. Plasma Sci.*, 13(5):242–252, 1985.

[10] H. C. Miller. Vacuum arc anode phenomena. *IEEE Trans. Plasma Sci.*, 5(3):181–196, 1977.

[11] H. C. Miller. Anode modes in vacuum arcs. *IEEE Trans. Dielectr. Elect. Insul.*, 4(4):382–388, 1997.

[12] H. C. Miller. Anode modes in vacuum arcs: Update. *IEEE Trans. Plasma Sci.*, 45(8):2366–2374, 2017.

[13] J. Janiszewski and Z. Zalucki. Photographic appearance of high-current vacuum arcs prior to and during anode spot formation. *Czechoslovak Journal of Physics.*, 46(10):961–971, 1996.

[14] Z. Zalucki. Voltage/current characteristics of high-current vacuum arc at a small gap length. *Czechoslovak J. Phys.*, 46(10):981–993, 1996.

[15] E. Dullni, E. Schade, and B. Gellert. Dielectric recovery of vacuum arcs after strong anode spot activity. *IEEE Trans. Plasma Sci.*, 15(5):538–544, 1987.

[16] R. L. Boxman. Measurement of anode surface temperature during a high current vacuum arc. *J. Appl. Phys.*, 46(11):4701, 1975.

[17] H. Rosenthal, I. Beilis, S. Goldsmith, and R. L. Boxman. A spectroscopic investigation of the development of a hot-anode vacuum arc. *J. Phys. D: Appl. Phys.*, 29(5):1245–1259, 1996.

[18] B. Gellert, E. Schade, and R. L. Boxman. Time- and spatially resolved spectroscopy of the plasma state prior to and during anode-spot formation in high-current vacuum arcs. *IEEE Trans. Plasma Sci.*, 13(5):265–268, 1985.

[19] H. Toya, Y. Uchida, T. Hayashi, and Y. Murai. Spectroscopic measurement and analysis of high-current vacuum arc near copper cathode. *IEEE Trans. Plasma Sci.*, 14(4):471–476, 1986.

[20] R. Methling, S. Gorchakov, M. V. Lisnyak, St. Franke, A. Khakpour, S. A. Popov, A. V. Batrakov, D. Uhrlandt, and K. D. Weltmann. Spectroscopic investigation of a Cu-Cr vacuum arc. *IEEE Trans. Plasma Sci.*, 43(8):2303–2309, 2015.

[21] M. V. Lisnyak, A. V. Pipa, S. Gorchakov, S. Iseni, St. Franke, A. Khakpour, R. Methling, and K. D. Weltmann. Overview spectra and axial distribution of spectral line intensities in a high-current vacuum arc with Cu-Cr electrodes. *J. Appl. Phys.*, 118(12):123304, 2015.

[22] B. Gellert. Measurement of high copper vapour densities by laser-induced fluorescence. *J. Phys. D: Appl. Phys.*, 22(5):710, 1988.

[23] B. Gellert, E. Schade, and E. Dullni. Measurement of particles and vapor density after high-current vacuum arcs by laser techniques. *IEEE Trans. Plasma Sci.*, 15(5):545–551, 1987.

[24] Z. Wang, H. Ma, G. Kong, Z. Liu, Y. Geng, and J. Wang. Decay modes of anode surface temperature after current zero in vacuum arcs-part I: Experimental study. *IEEE Trans. Plasma Sci.*, 42(5):1462–1473, 2014.

[25] Z. Wang, Y. Tian, H. Ma, Y. Geng, and Z. Liu. Decay modes of anode surface temperature after current zero in vacuum arcs—part II: Theoretical study of dielectric recovery strength. *IEEE Trans. Plasma Sci.*, 43(10):3734–3743, 2015.

[26] R. Methling, St. Franke, S. Gortschakow, M. Abplanalp, R. P. Sütterlin, T. Delachaux, and K. O. Menzel. Anode surface temperature determination in high-current vacuum arcs by different methods. *IEEE Trans. Plasma Sci.*, 45(8):2099–2107, 2017.

[27] S. W. Rowe. The intrinsic limits of vacuum interruption. In *Proc. 23rd Int. Symp. Discharges Elect. Insul. Vac.*, pages 192–197, Bucharest, Romania, 2008.

[28] The impact of the application of vacuum switchgear at transmission voltages. In *CIGRE working group A3.14 internal document A3-14 (SC) 02 IWD.*, 2014.

[29] IEEE application guide for transient recovery voltage for ac high-voltage circuit breakers. In *IEEE Std C37.011*, New York 10016-5997, USA, 2005.

[30] S. M. Shkol'nik. Secondary plasma in the gap of high-current vacuum arc: Origin and resulting effects. *IEEE Trans. Plasma Sci.*, 31(5):832–846, 2003.

[31] W. Li, R. L. Thomas, and R. K. Smith. Effects of Cr content on the interruption ability of Cu-Cr contact materials. *IEEE Trans. Plasma Sci.*, 29(5):744–748, 2001.

[32] M. Sugita, S. Okabe, G. Ueta, W. Wang, X. Wang, and S. Yanabu. Interruption phenomena for various contact materials in vacuum. *IEEE Trans. Plasma Sci.*, 37(8):1469–1476, 2009.

[33] G. Kong, Z. Liu, D. Wang, and M. Rong. High-current vacuum arc: The relationship between anode phenomena and the average opening velocity of vacuum interrupters. *IEEE Trans. Plasma Sci.*, 39(6):1370–1378, 2011.

[34] G. A. Dyuzhev, A. Lyubimov, and S. M. Shkol'nik. Conditions of the anode spot formation in a vacuum arc. *IEEE Trans. Plasma Sci.*, 11(1):36–45, 1983.

[35] L. Wang, J. Deng, H. Wang, S. Jia, K. Qin, and Z. Shi. Vacuum arc behavior and its voltage characteristics in drawing process controlled by composite magnetic fields along axial and transverse directions. *J. Phys. Plasma*, 22(10):103512, 2015.

[36] C. W. Kimblin. Anode voltage drop and anode spot formation in dc vacuum arc. *J. Appl. Phys.*, 40(4):1744–1752, 1969.

[37] S. Agarwal and R. Holmes. Arcing voltage of the metal vapor vacuum. *J. Phys. D: Appl. Phys.*, 17(4):757–767, 1984.

[38] S. A. Popov, A. V. Schneider, V. Lavrinovich, A. V. Batrakov, S. Gortschakow, and A. Khakpour. Fast video registration of transition processes from diffuse mode to anode spot mode in high-current arc with copper-chromium electrodes. In *Proc. 27th Int. Symp. Discharges Elect. Insul. Vac.*, pages 375–378, Suzhou, China, 2016.

[39] L. Sun, L. Yu, Z. Liu, J. Wang, and Y. Geng. An opening displacement curve characteristic determined by high-current anode phenomena of a vacuum interrupter. *IEEE Trans. Power Del.*, 28(4):2585–2593, 2013.

[40] T. Shioiri, T. Kamikawaji, E. Kaneko, M. Homma, H. Takahashi, and I. Ohshima. Influence of electrode area on conditioning in vacuum gap breakdown. *IEEE Trans. Dielectr. Elect. Insul.*, 2(2):317–320, 1995.

[41] A. V. Batrakov, A. V. Schneider, S. Rowe, G. Sandolache, A. Markov, and L. Zjulkova. Observation of an anode spot shell at the high-current vacuum arc. In *Proc. 24th Int. Symp. Discharges Elect. Insul. Vac.*, pages 351–354, Braunschweig, Germany, 2010.

[42] A. V. Schneider, S. A. Popov, G. Sandolache, A. V. Batrakov, and S. Rowe. Ionization–recombination front in high-current vacuum arc. *IEEE Trans. Plasma Sci.*, 39(11):2844–2845, 2011.

[43] A. V. Batrakov, S. V. Popov, A. V. Schneider, G. Sandolache, and S. W. Rowe. Observation of the plasma plume at the anode of high-current vacuum arc. *IEEE Trans. Plasma Sci.*, 39(6):1291–1295, 2011.

[44] R. Methling, S. Gorchakov, M. V. Lisnyak, St. Franke, , A. Khakpour, S. A. Popov, A. V. Batrakov, D. Uhrlandt, and K. D. Weltmann. Spectroscopic investigation of high-current vacuum arcs. In *Proc. 27th Int. Symp. Discharges Elect. Insul. Vac.*, pages 221–224, Mumbai, India, 2014.

[45] I. I. Beilis. State of the theory of vacuum arcs. *IEEE Trans. Plasma Sci.*, 29(5):657–670, 2001.

[46] L. Wang, X. Zhou, H. Wang, Z. Qian, S. Jia, D. Yang, and Z. Shi. Anode activity in a high-current vacuum arc: Three-dimensional modeling and simulation. *IEEE Trans. Plasma Sci.*, 40(9):2234–2246, 2012.

[47] L. Wang, X. Huang, S. Jia, J. Deng, Z. Qian, Z. Shi, H. Schellenkens, and X. Godechot. 3D numerical simulation of high current vacuum arc in realistic magnetic fields considering anode evaporation. *J. Appl. Phys.*, 117(24):243301, 2015.

[48] L. Zhang, L. Wang, S. Jia, D. Yang, and Z. Shi. Numerical simulation of high-current vacuum arc with consideration of anode vapor. *Plasma Sci. Technol.*, 14(4):285–292, 2012.

[49] X. Huang, L. Wang, X. Zhang, S. Jia S, and Z. Shi. Numerical simulation of HCVA with considering the micro process of anode vapor. In *Proc. 27th Int. Symp. Discharges Elect. Insul. Vac.*, pages 263–266, Suzhou, China, 2016.

[50] O. Mayr. Beiträge zur Theorie des statischen und des dynamischen Lichtbogens. *Archiv Elektrotech.*, 37(12):588–608, 1943.

[51] M. Cassie. Theorie nouvelle des arcs de rupture et de la rigidité des circuits. *Cigre*, 102:588–608, 1939.

[52] U. Habedank. Application of a new arc model for the evaluation of short-circuit breaking tests. *IEEE Trans. Power Del.*, 8(4):1921–1925, 1993.

[53] J. A. Martinez-Velasco. *Power System Transients: Parameter Determination*. CRC Press, Boca Raton, 2010.

[54] W. Gimenez and O. Hevia. Method to determine the parameters of the electric arc from test data. In *Proc. Int. Conf. Power Systems Transient*, pages 505–509, Budapest, Hungary, 1999.

[55] M. M. Walter and C. M. Franck. Improved method for direct black-box arc parameter determination and model validation. *IEEE Trans. Power Del.*, 29(2):580–588, 2014.

[56] J. A. Martinez, J. Mahseredjian, and B. Khodabakhchian. Parameter determination for modeling system transients-part VI: Circuit breakers. *IEEE Trans. Power Del.*, 20(3):2079–2085, 2005.

[57] A. Sawicki. Problems of modeling an electrical arc with variable geometric dimensions. *Przeglad Elektrotechn.*, 2013(2b):270–275, 2013.

[58] P. H. Schavemaker and L. van der Sluis. An improved Mayr-type arc model based on current-zero measurements. *IEEE Trans. Power Del.*, 15(2):580–584, 2000.

[59] S. Nitu, C. Nitu, C. Mihalache, P. Anghelita, and D. Pavelescu. Comparison between model and experiment in studying the electric arc. *J. Optoelectron. Adv. Mater.*, 10(5):1192–1196, 2008.

[60] T. Koshizuka, K. Udagawa, T. Iijima, T. Uchii, T. Shinkai, and T. Mori. Investigation of arc parameters in serially-connected 3 arc model. In *Conf. Power Systems Transients*, pages 56–61, Vancouver, Canada, 2013.

[61] D. Uhrlandt, R. Methling, St. Franke, S. Gorchakov, M. Baeva, St. Franke, A. Khakpour, and V. Brueser. Extended methods of emission spectroscopy for the analysis of arc dynamics and arc interaction with walls. *J. Plasma Phys. Tech.*, 2(3):280–289, 2015.

[62] N. Gustavsson. Evaluation and simulation of black-box arc models for high voltage circuit-breakers. Master's thesis, Dept. Elect. Eng., Linköping Univ., Sweden, 2004.

[63] A. Khakpour, St. Franke, D. Uhrlandt, S. Gorchakov, and R. Methling. Electrical arc model based on physical parameters and power calculation. *IEEE Trans. Plasma Sci.*, 43(8):2721–2729, 2015.

[64] A. Khakpour, D. Uhrlandt, R. Methling, S. Gortschakow, St. Franke, M. T. Imani, and K. D. Weltmann. Impact of temperature changing on voltage and power of an electric arc. *Electr. Power Sys. Res.*, 143:73–83, 2017.

[65] Y. Cressault and A. Gleizes. Thermal plasma properties for Ar–Al, Ar–Fe and Ar–Cu mixtures used in welding plasmas processes: I. Net emission coefficients at atmospheric pressure. *J. Phys. D: Appl. Phys.*, 46(41):415206, 2013.

[66] Y. Cressault, A. B. Murphy, Ph. Teulet, A. Gleizes, and M. Schnick. Thermal plasma properties for Ar–Al, Ar–Fe and Ar–Cu mixtures used in welding plasmas processes: II. Transport coefficients at atmospheric pressure. *J. Phys. D: Appl. Phys.*, 46(41):415207, 2013.

[67] D. Uhrland, S. Gorchakov, V. Brueser, St. Franke, A. Khakpour, M. Lisnayek, R. Methling, and Th. Schoenemann. Interaction of a free burning arc with regenerative protective layers. *J. Phys.: Conf. Ser.*, 550(1):012010, 2014.

[68] K. C. Hsu and E. Pfender. Modeling of a free-burning, high-intensity arc at elevated pressures. *Plasma Chem. Plasma Proc.*, 4(3):219–234, 1984.

[69] A. Khakpour, St. Franke, S. Gortschakow, D. Uhrlandt, R. Methling, and K. D. Weltmann. An improved arc model based on the arc diameter. *IEEE Trans. Power Del.*, 31(3):1335–1341, 2016.

[70] J. J. Lowke and H. C. Ludwig. A simple model for high-current arcs stabilized by forced convection. *J. Appl. Phys.*, 46(8):3352–3360, 1975.

[71] S. Berger. *Modell zur Berechnung des dynamischen elektrischen Verhaltens rasch verlängerter Lichtbögen*. PhD thesis, ETH Zürich, Switzerland, 2009.

[72] M. Walter, M. Kang, and C. M. Franck. Arc cross-section determination of convection stabilized arcs. In *18th Int. Symp. High Vol. Eng.*, pages 164–69, Seoul, South Korea, 2016.

[73] A. Khakpour, St. Franke, R. Methling, D. Uhrlandt, S. Gortschakow, S. A. Popov, A. V. Batrakov, and K. D. Weltmann. Optical and electrical investigation of transition from anode spot type 1 to anode spot type 2. *IEEE Trans. Plasma Sci.*, 45(8):2126–2134, 2017.

[74] B. Horvath and T. Lamara. Time-resolved optical resonant absorption spectroscopy of Cr metallic vapor in air using a broadband led light source. *Plasma Sources Sci. Technol.*, 22(4):045006, 2013.

[75] St. Franke, R. Kozakov, S. Gortschakow, A. Khakpour, R. Methling, and D. Uhrlandt. Broadband absorption technique applied to a free-burning arc in ambient air. In *21th Int. Conf. on Gas Discharges and their Application*, pages 5–8, Nagoya, Japan, 2016.

[76] A. Khakpour, S. A. Popov, St. Franke, R. Kozakov, R. Methling, D. Uhrlandt, and S. Gortschakow. Determination of Cr density after current zero in a high-current vacuum arc considering anode plume. *IEEE Trans. Plasma Sci.*, 45(8):2108–2114, 2017.

[77] S. Gortschakow, A. Khakpour, S. A. Popov, St. Franke, R. Methling, and D. Uhrlandt. Determination of Cr density in a high-current vacuum arc considering anode activity. *J. Plasma Phys. Tech.*, 4(3):190–193, 2017.

[78] K. Günther and R. Radtke. A proposed radiation standard for the visible and UV region. *J. Phys. E: Sci. Inst.*, 8(5):371–376, 1975.

[79] St. Franke, R. Methling, D. Uhrlandt, R. Bianchetti, R. Gati, and M. Schwinne. Temperature determination in copper-dominated free-burning arcs. *J. Phys. D: Appl. Phys.*, 45(1):015202, 2014.

[80] J. Zalach and St. Franke. Iterative Boltzmann plot method for temperature and pressure determination in a xenon high pressure discharge lamp. *J. Appl. Phys.*, 113(4):043303, 2013.

[81] D. Uhrlandt. Diagnostics of metal inert gas and metal active gas welding processes. *J. Phys. D: Appl. Phys.*, 49(31):313001, 2016.

[82] A. Khakpour, R. Methling, D. Uhrlandt, St. Franke, S. Gortschakow, S. A. Popov, A. V. Batrakov, and K. D. Weltmann. Time and space resolved spectroscopic investigation during anode plume formation in high-current vacuum arc. *J. Phys. D: Appl. Phys.*, 50(18):185203, 2017.

[83] A. N. Zaidel, V. K. Prokof'ev, and S. M. Raiskii. *Tables of Spectrum Lines*. New York, NY, USA: Plenum, 1970.

[84] A. Kramida, Y. Ralchenko, and J. Reader. NIST atomic spectra database, 2011. [online] http://physics.nist.gov/asd.

[85] R. L. Kurucz and B. Bell. Atomic line data. kurucz, 1995. [online]http://www.cfa.harvard.edu/amp/ampdata/kurucz23/sekur.html.

[86] A. G. Shenstone. The third spectrum of copper (Cu III). *J. Res. Nat. Bur. Stand.*, 79A(3):497–521, 1975.

[87] N. Konjevic, A. Lesage, J. R. Fuhr, and W. L. Wiese. Experimental stark widths and shifts for spectral lines of neutral and ionized atoms (a critical review of selected data for the period 1983 through 2000). *J. Phys. Chem. Ref. Data*, 31(3):819–927, 2002.

[88] M. Skočić, M. Burger, Z. Nikolić, S. Bukvić, and S. Djeniže. Stark broadening in the laser-induced Cu I and Cu II spectra. *J. Phys. B: Mol. Opt. Phys*, 46(18):185701, 2013.

[89] B. Zmerli, N. Ben Nessib, M. S. Dimitrijević, and S. Sahal-Bréchot. Stark broadening calculations of neutral copper spectral lines and temperature dependence. *Phys. Scr.*, 82(5):055304, 2010.

[90] L. Yang, X. Tan, X. Wan, L. Chen, D. Jin, M. Qian, and G. Li. Stark broadening for diagnostics of the electron density in non-equilibrium plasma utilizing isotope hydrogen alpha lines. *J. Appl. Phys.*, 115(16):163106, 2014.

[91] A. Khakpour. Impact of AC and pulsed DC interrupting currents on the formation of high-current anode modes in vacuum. In *Proc. 27th Int. Symp. Discharges Elect. Insul. Vac.*, pages 129–132, Suzhou, China, 2016.

[92] A. Khakpour, D. Uhrlandt, R. Methling, St. Franke, S. Gortschakow, S. A. Popov, A. V. Batrakov, and K. D. Weltmannn. Impact of different vacuum interrupter properties on high-current anode phenomena. *IEEE Trans. Plasma Sci.*, 44(12):3337–3345, 2016.

[93] A. Khakpour, S. Gortschakow, S. A. Popov, R. Methling, St. Franke, D. Uhrlandt, A. V. Batrakov, and K. D. Weltmann. Time and space resolved video spectroscopy of the vacuum arc during the formation of high-current anode mode. In *Proc. 27th Int. Symp. Discharges Elect. Insul. Vac.*, pages 287–290, Suzhou, China, 2016.

[94] A. Khakpour, S. Gortschakow, D. Uhrlandt, R. Methling, St. Franke, S. A. Popov, A. V. Batrakov, and K. D. Weltmann. Video spectroscopy of vacuum arcs during transition between different high-current anode modes. *IEEE Trans. Plasma Sci.*, 44(10):2462–2469, 2016.

[95] A. Khakpour, R. Methling, St. Franke, S. Gortschakow, and D. Uhrlandt. Emission spectroscopy during high-current anode modes in vacuum arc. *J. Plasma Phys. Tech.*, 4(4), 2017.

[96] J. V. R. Heberlein and J. G. Gorman. The high current metal vapor arc column between separating electrodes. *IEEE Trans. Plasma Sci.*, 8(4):283–288, 1980.

[97] G. Bizjak, P. Zunko, and D. Povh. Combined model of SF6 circuit breaker for use in digital simulation programs. *IEEE Trans. Power Del.*, 19(1):174–180, 2004.

[98] J. L. Guardado, S. G. Maximov, E. Melgoza, J. L. Naredo, and P. Moreno. An improved arc model before current zero based on the combined Mayr and cassie arc models. *IEEE Trans. Power Del.*, 20(1):138–142, 2005.

[99] A. Ahmethodžić, M. Kapetanović, K. Sokolija, R. P. P. Smeets, and V. Kertész. Linking a physical arc model with a black box arc model and verification. *IEEE Trans. Dielectr. Electr. Insul.*, 18(4):1029–1037, 2011.

[100] H. Wu, L. Yuan, L. Sun, and X. Li. Modeling of current-limiting circuit breakers for the calculation of short-circuit current. *IEEE Trans. Power Del.*, 30(2):652–656, 2015.

[101] H. A. Darwish and N. I. Elkalashy. Universal arc representation using EMTP. *IEEE Trans. Power Del.*, 20(2):772–779, 2005.

[102] A. Khakpour, M. T. Imani, R. Methling, St. Franke, S. Gortschakow, and D. Uhrlandt. An improved electric arc model for vacuum arc regarding anode spot modes. *Electr. Power Sys. Res.*, submitted, 2017.

[103] M. B. J. Leusenkamp. Vacuum interrupter model based on breaking test. *IEEE Trans. Plasma Sci.*, 27(4):969–976, 1999.

[104] G. Kong, Z. Liu, Y. Geng, H. Ma, and X. Xue. Influence of contact solid angle on anode spot formation threshold current in vacuum circuit breakers. In *Proc. 25th Int. Symp. Discharges Elect. Insul. Vac.*, pages 333–336, Tomsk, Russia, 2012.

[105] G. Kong, Z. Liu, Y. Geng, H. Ma, and X. Xue. Anode spot formation threshold current dependent on dynamic solid angle in vacuum subjected to axial magnetic fields. *IEEE Trans. Plasma Sci.*, 41(8):2051–2060, 2013.

# D. Curriculum Vitae

## Personal information

| | |
|---|---|
| **Name:** | Alireza Khakpour |
| **Birthday:** | 10.09.1981 |
| **Email:** | as.khakpour@gmail.com |
| **Marital status:** | Married |

## Academic and work background

| | |
|---|---|
| **11.2017 to present** | Senior Consultant<br>ABB Power Consulting Germany |
| **09.2013 to 11.2017** | Staff scientist(PhD candidate)<br>Leibniz Institute for Plasma Science and Technology (INP Greifswald) |
| **06.2011 to 09.2013** | Ghods Niroo Consultant Engineering (GNCE) as head of Power System Study Department |
| **03.2009 to 06.2011** | M.Sc. electrical engineering, Joint Program at Leibniz Universität Hannover, Germany and K. N. Toosi University of Teheran, Iran |
| **06.2004 to 03.2009** | Ghods Niroo Consultant Engineering as<br>Junior and Senior Engineer<br>in Power System Study Department |
| **09.1999 to 06.2004** | B.Sc. electrical engineering at<br>K. N. Toosi University of Teheran, Iran |
| **09.1994 to 09.1999** | High school in Mathematics and Physics at<br>Palayeshgah High School, Abadan, Iran |